NATIONAL NUMERACY TESTS Y7

A+ National Practice Tests
graduated difficulty with solutions

**Sue Ferguson
Sarah Hamper
Wendy Bodey**

A+ National

A+ National Numeracy Tests Year 7
1st Edition
Sue Ferguson
Sarah Hamper
Wendy Bodey

Publishing editors: Jane Moylan and Jana Raus
Editor: Kerry Nagle
Project packager: UC Publishing
Senior designer: Ami Sharpe
Text designer: Ami Sharpe
Cover designer: Ami Sharpe
Cover image: Getty Images
Photo researcher: UC Publishing/Shelley Underwood
Production controller: Jo Vraca and Damian Almeida
Reprint: Katie McCappin
Typeset by UC Publishing Pty Ltd

Any URLs contained in this publication were checked for currency during the production process. Note, however, that the publisher cannot vouch for the ongoing currency of URLs.

Acknowledgements
VCE ® is a registered trademark of the VCAA. The VCAA does not endorse or make any warranties regarding this Cengage product. Current and past VCE Study Designs, VCE exams and related content can be accessed directly at www.vcaa.vic.edu.au

We would like to thank the following people for reviewing this title: Howard Reeves, Stephen Swift, Stephen Corcoran and Janet Hunter.

We would like to thank the following for permission to reproduce copyright material:

iStockphoto: p46 track5.

For product information and technology assistance,
in Australia call **1300 790 853**;
in New Zealand call **0800 449 725**

For permission to use material from this text or product, please email **aust.permissions@cengage.com**

ISBN 978 0 17 046220 4

Cengage Learning Australia
Level 7, 80 Dorcas Street
South Melbourne, Victoria Australia 3205

Cengage Learning New Zealand
Unit 4B Rosedale Office Park
331 Rosedale Road, Albany, North Shore 0632, NZ

For learning solutions, visit **cengage.com.au**

Printed in China by 1010 Printing International Limited.
3 4 5 6 7 25 24 23 22

Detailed Information

Introduction

▶ To the student

Literacy and numeracy are the fundamental building blocks of learning in any subject. Knowing what you can do and where you need to improve is vital for all Australian students, teachers and parents. The NAPLAN* (National Assessment Program – Literacy and Numeracy) tests in literacy and numeracy help governments find out how Australian students are progressing and help to identify what you know and what you don't know. The results of the tests also help you and your teachers plan what you need to learn next.

Tests can sometimes be a little daunting; however, there are practical steps you can take to ensure you will successfully sit a test. You will be best prepared for any test when you understand what is being tested, how you will be tested, and when you are mentally and physically prepared for the test. This book provides practical advice and strategies to ensure that you are test ready and contains practice tests so you can see what to expect in the tests.

This book and the accompanying NelsonNet website provide:

✔ **Test tips:** advice on how to successfully sit tests, information about what is tested and how it is tested, hints on responding to different question types and how to act on your results.

✔ **Calculator skills:** provides a clear tutorial to ensure you make the best use of your calculator.

✔ **Comprehensive practice tests:** includes three non-calculator practice tests and three calculator-allowed practice tests. These tests are graduated in difficulty and content so that you can build your skills and confidence during the first part of the year leading up to the test.

✔ **Four full-length tests:** four full-length detachable tests (two non-calculator and two calculator-allowed tests), one set is for you to use as a practice test, and the other set is for you to hand in to your teacher. The full-length tests are of the same length

and level of difficulty you can expect to find when you sit the NAPLAN* tests so are ideal for practise during the month leading up to the test.

✔ **Answers:** for easy reference, answers to all workbook questions can be found on NelsonNet, https://www.nelsonnet.com.au/free-resources.

✔ **Tips for teachers and handy checklists:** for easy reference teacher hints and checklists are provided on the NelsonNet website.

✔ **Useful icons:** icons are used throughout the resources to help you quickly find other important information in the resources, support your understanding and clearly distinguish questions where you may use a calculator if you wish.

✔ You will also notice the following key features throughout this book:

Hot tips: keep one step ahead by heeding these hot tips. The tips often relate to things many students forget to do or can do better with a little planning.

World wide web link: provides a useful link to further information available on the Internet.

Calculator-allowed: this indicates you may use a calculator to assist you to answer the question and when you check your work.

Non-calculator: this indicates you must complete any calculations without the use of a calculator.

About the authors

Sue Ferguson

Sue Ferguson has worked in mathematics and curriculum and assessment for many years. Presently Sue lectures at Victoria University, and prior to that she worked at the Australian Curriculum, Assessment and Reporting Authority (ACARA) heading a team of writers to develop the Australian Mathematics Curriculum. She has also worked at the Curriculum Corporation on The Le@rning Federation project where she provided mathematics advice on the development of digital curriculum resources. Sue is also an experienced maths teacher and has been a secondary mathematics teacher at all levels in government schools in Victoria.

Wendy Bodey

Wendy Bodey draws on diverse experience gained working at Curriculum Corporation, the Australian Council for Educational Research and as a teacher. Wendy has authored a variety of assessment materials and resources and has played key roles in national and jurisdiction assessment programs including the inaugural National Assessment Program Literacy and Numeracy (NAPLAN) tests.

Sarah Hamper

Sarah Hamper is an experienced mathematics teacher and currently teaches in New South Wales at Abbotsleigh School. Sarah has a Bachelor degree in Pure Mathematics and a Master's degree in Education, studying mathematical modelling, problem solving and using real world applications and Information Communication Learning Technologies (ICLT) to aid effective teaching and learning of mathematics. Sarah has also presented a number of professional papers in this area at state and national level.

Test tips

All tests are designed to provide useful information to help you and your teacher plan what you need to learn next. The NAPLAN* tests are no different. All you need to do is be as prepared as you can and answer the questions to the best of your ability. Follow the practical advice provided throughout the rich resources in this book and the accompanying NelsonNet wesbite and you will be prepared to successfully sit most tests at the secondary level.

Generally the questions become more difficult as you proceed through the test

The numeracy test questions cover a range of content and difficulty levels. Understanding more about the types of skills assessed at each level is useful information for you as a test taker.

To illustrate the range of difficulty in a test think about the following questions you might be asked to answer about information in tables. Early in the test you might simply be asked to locate particular information contained in a table, such as the number of students who ride to school. By the middle of the test you might be asked to interpret information in a table, such as working out the best time to catch a train to arrive at your destination by 4 p.m. Towards the end of the test you might be asked to calculate and interpret the mean, median and mode of data in a table. The test questions vary from year to year, however, the range of difficulty and content across the test will be very similar.

Use the icons

The majority of test questions contained in the two booklets are multiple-choice format while approximately one-quarter of the test questions require you to provide the correct answer. There are icons on the page that tell you what to do, for example, 'Shade one bubble' or 'Write your answer in the box'. Follow the icon instructions as you proceed through the test.

There are two separate numeracy tests. You will be able to use a calculator to assist you in one of the test papers so make sure you read the calculator skills section to ensure you can make the best use of your calculator during that part of the test. Throughout these resources distinctive icons will help you identify when you are allowed to use a calculator.

There are icons to alert you to further information available on the Internet. Make the most of this extra information to keep you one step ahead.

And of course, make sure you read all the special tips throughout the resources that are highlighted by the 'Hot tips' icon as these tips relate to things many students forget to do or can do better.

Read the questions carefully

The first step is to read the question and think about what it is asking you to do. Try underlining key information in the question to focus your thinking. Consider the following question and then read the recommended approach that you can apply to any question.

QUESTION 4 WRITE YOUR OWN ANSWER

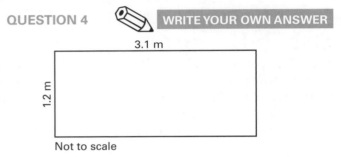

Not to scale

What is the area of this rectangle? Round your answer to the nearest square metre.

It is useful to circle or underline the measurements given on the rectangle: 3.1 metres (note this is the longer side, or the length), 1.2 metres (note this is the shorter side, or the width).

The question text tells you what you need to calculate and the accuracy required for your answer. Underline the important words in the question such as: area, rectangle, nearest square metre.

The icon indicates that you are required to write your answer in the box provided. It is important that your answer is complete. In this case the answer needs to be correct to the nearest square metre.

To answer this question you apply the formulae for calculating the area of a rectangle. The area of a rectangle is the length times the width.

$$A = l \times w \text{ where } l = \text{length and } w = \text{width}$$

Remember to revise important facts and formulae such as this during your test preparation.

The area of this rectangle is 3.1 × 1.2. Your calculation is 3.1 × 1.2 or a total of 3.72 square metres. Rounded to the nearest square metre the area is 4 square metres. It is very important that you include the measurement unit (square metres or m²) to ensure the answer is completely accurate. Likewise, if a question has any unit description this should be included in your answer, for example 28 US dollars, 115 degrees,

---

10 hours, 49%. Sometimes the unit will already be provided.

Check your answer

It is important to always check your answer! Try using a different method to calculate the answer, if this is possible, as this is a good way to cross check your answer. Estimating the answer is a good way of checking. To estimate the area you could round each number. In this instance 3.1 is rounded to 3 and 1.2 is rounded to 1. The estimate of the area is 3 × 1 or 3 square metres. This gives you an approximate check against your calculation of 3.72 square metres.

Sometimes the question will require you to think through the practical application of your calculations. Imagine the question asked you: 'Tom wants to cover a rectangular room 3.1 metres long and 1.2 metres wide with floor tiles. Each tile is one hundred centimetres square. How many floor tiles will he need?'

In this case you would need to ensure that the quantity of floor tiles is adequate to completely cover the floor, otherwise Tom will have a section of floor not covered. You would need to round the area to one decimal place to ensure you have the correct number of tiles. You round

3.72 to 3.8 square metres. Tom will need 38 floor tiles.

Eliminate incorrect options in multiple-choice questions

Multiple-choice questions provide the correct answer along with some other, often attractive, options for you to choose from. It is important therefore that you carefully do your own calculations before reviewing the answer options. In this way you will not be distracted by attractive, but incorrect, options.

It is a myth that 'B' or any other option is usually correct. Most tests have a good mix of correct answers A, B, C or D, randomly spread across the test with approximately the same number of As, Bs and the other options available. Some students think if they are not sure of an answer they should stick with a particular guess, for example always choosing 'B' when unsure. This is not a recommended strategy as it is unlikely to be successful and it is also not a useful learning strategy.

Likewise, don't be fooled into thinking that just because two options are similar to one another that one of these options must be the correct answer. In many instances neither of these options will be the correct answer.

When you are unsure which option is the correct answer start by trying to eliminate any obviously incorrect options. If a question asked you to calculate 53 × 26 then it would be easy to eliminate the option of 79 as a potential answer as this is the sum of the two numbers rather than its product. Multiplying the two units (3 and 6) equals 18, and multiplying the tens (5 and 2) equals 10; therefore 1018 might initially look like an attractive option. However, this is not the correct way to calculate 53 multiplied by 26. We can get a good sense of the correct answer by rounding 53 × 26 to their nearest ten (50 × 30) and see the answer will be approximately 1500. Therefore 1018 can clearly be eliminated as too low an answer. Use strategies such as these to eliminate options and also assist you to confirm the correct answer.

Watch the time but don't hurry

The time available to complete each of the tests is adequate for most students to comfortably complete the test. So be conscious of the time as you work through each test but don't rush your answers.

You will have 40 minutes to complete each test book; this is the equivalent of just over one minute per question. Start by working through the questions you are most confident to answer and do quick checks as you go. Use the remaining test time to tackle questions you are less sure of and to more thoroughly check your answers. It is not wise to spend too much time on any one question until you have completed all other questions in the test.

Prepare well

Some students feel nervous when it comes to test time. The best way to manage this is by ensuring you are well prepared before test day, and then you will have no need for any concern. Find out as much as you can about the tests well before you sit them so that you have an understanding of what is being tested, and how it will be tested. Read through the calculator skills section, revise important facts and formulae and review mathematical terminology to make sure you recognise and understand common terms. You will find this revision useful during the test and in building your knowledge.

Work through the three non-calculator and three calculator-allowed practice tests provided to give you a sense of what the test will be like and to assist you to identify areas that you may need to revise. It is also a good idea to check how efficient you are with your use of time. Consider any action you may take to improve this efficiency.

Then complete the four full-length detachable tests (two non-calculator and two calculator-allowed). The full-length tests are of the same length and level of difficulty you can expect to find when you sit the actual tests. The first set is designed for you to use as a practice test, the other set to hand in to your teacher. Discuss your test results with your teacher and seek their feedback on how well you performed and any areas you might improve on.

Be calm and don't panic

On test day make sure you have had a good night's sleep and that you have eaten breakfast so that you are physically prepared to do the test. Be confident in your preparation and you will be ready to tackle the test. Use the pre-test and test day checklists provided to assist your preparation.

It is important that you answer each test question to the best of your ability as this will provide a better guide for your future learning requirements. Tests help to identify your strengths and any potential areas for further development that you may have. It is important that you discuss your results with your teacher to ensure your learning program and goals are the most appropriate for you.

Enjoy adding to your skills

Include mathematics activities in your daily routine and you will be surprised how enjoyable it is to further build your skills. Try solving at least one puzzle every day. You will find puzzles in daily newspapers, magazines and there are many computer software games that include mathematical puzzles. Solving a Sudoku each day, for example, will enhance your problem solving skills.

Take note of maths you see during the day such as the speed you travel in a car, the minimum and maximum daily temperature, distance travelled to a destination, the time it takes to complete a particular activity and compare product prices, features and any discounts available to work out which brand is the best value.

Study tables, graphs and other data to build your data reading and interpreting skills such as making sure you can identify the country with the highest population or that you can interpret the speed of a jet during different periods of its flight.

Maths is all around you. Start taking more notice and you will build your knowledge and your skills

and at the same time you will become more proficient in the application of mathematics in your day-to-day life.

Know the terms, facts and formula

It is important to know mathematical facts such as the names of shapes and solids – do you know the difference between an octagon and a hexagon or the difference between a sphere and a cylinder? Facts such as these are integral to you being able to understand questions posed in the test and to solve problems and apply your reasoning to calculations.

Well before test day revise common mathematical facts and make sure you understand the terms used in maths. Start by familiarising yourself with mathematical terms, for example 'mode' and 'median'. Make sure you know what they mean and where you are likely to use them. Make a point of adding new facts and terms to your current knowledge.

Make sure you have a good knowledge of common formulae such as how to calculate the perimeter of two dimensional shapes or the volume of a three dimensional shape.

Understand common conventions used in mathematics and how to apply these, such as how to read and interpret the use of indices. For example, knowing that 2^3 is the equivalent of $2 \times 2 \times 2$ or 8.

Be confident in comparing fractional amounts and being able to work with fractions, such as being able to reduce a fraction to its lowest terms or determine which fraction is the largest when comparing two or more fractions. This includes understanding decimal fractions and how the position of the decimal point indicates relative value. Start by exploring money as this is an everyday application of decimals, adding and subtracting quantities of money, calculating interest or a discount or planning your budget and saving goals. Then extend this into understanding the use of other decimal numbers such as measuring length, knowing that 0.3 km is the same as 300 m.

Knowing basic maths facts and the common formula and terms used will mean you do not waste time trying to understand what a question is asking you and will maximise the time available to solve the problem.

Answer position is important

Where you write your answer can impact on your results. Most of the questions you answer will be machine-scored so it is important your answer is in the correct place so the machine records your responses accordingly.

When responding to multiple-choice questions the instruction icon will say, 'Shade one bubble'. You need to identify the correct answer, this will sometimes be an illustration, and then colour the bubble that indicates this answer. Don't make the mistake of circling the correct illustration, or marking some information on a graph but forgetting to colour the appropriate bubble as you may not get credited for your response.

Make sure you only colour one bubble for each answer and that you colour it in completely. Don't just draw a line through the bubble as the machine may not detect this. If you want to change your answer make sure you erase your other answer completely so that the machine does not accidentally record more answers than you meant to provide. Typically you will not be credited if it appears you have more than one answer for any question in the tests.

Some questions require you to write your answer, this might be numbers, words, symbols or equations. Make sure your writing is legible and that your numbers are very clear, for example make sure your 3 doesn't look like an 8. If it is difficult to read your answer you may not get credited for a correct response. And make sure you write your answer in the box or space indicated or an assumption may be made that you did not answer that particular question.

Keep up to date with information from your test authority

It is important to keep up with any information about the test. Your test authority will provide regular updates. Contact details for all Australian Test Administration Authorities for the NAPLAN* tests can be found at:

www.naplan.edu.au/test_administration_authorities.html

Other general information about the tests and specific information such as test dates can be found at:

www.naplan.edu.au

Pre-test checklist

Use this checklist to ensure you are prepared to successfully sit the tests.

	Activity	Main Resource
☐	Revise important facts, formulae and terms used in numeracy and be familiar with common conventions used	Class texts and notes
☐	Know how to use your calculator	Calculator skills
☐	Build your skills and knowledge informally, too	Solve puzzles and play games Take notice of everyday maths
☐	Complete the six practice tests to get a clear idea about the tests and questions types	Complete the six practice tests
☐	Check your answers against the solutions to evaluate your strengths and any areas you need to revise	Practice test solutions
☐	Get advice on tackling the tests	Test tips
☐	Complete Full-length Tests 7 and 8, the same length and level of difficulty you can expect in the test	Full-length Tests 7 and 8
☐	Check your answers against the solutions to evaluate your strengths and any areas you need to revise	Full-length Tests 7 and 8 solutions
☐	Complete Full-length Tests 9 and 10 and hand in to your teacher	Full-length Tests 9 and 10
☐	Ask your teacher for feedback on your performance and use of time	You and your teacher
☐	Keep updated on test information from your assessment authority	Test Administration Authority

Test day checklist

Use this checklist to ensure you are prepared on test day

Day before the test

	Activity
☐	Calculator, HB or 2B pencils, an eraser and a sharpener ready to take
☐	Have a good night's sleep

Test morning

	Activity
☐	Have breakfast
☐	Take watch, calculator, pencils, eraser and sharpener
☐	Arrive at school, or the testing venue, well before the session commences

During the test

	Activity
☐	Be confident in your preparation
☐	Monitor your time during the test
☐	Work through the test, completing easy questions first
☐	Read each question carefully, underline important words
☐	Don't spend too much time on any one question
☐	Circle the question number of any question you need to return to
☐	Make sure your written answers are legible
☐	Only write in the box or on the lines provided
☐	Select the correct option in a multiple-choice question, check the other options are incorrect
☐	Choose one option only and colour in the bubble or box completely
☐	If you change your answer, rub out the other answer completely
☐	Go back to complete unanswered questions
☐	Wait a question or two before going back to check your answer
☐	Check your work, make sure you haven't skipped any questions

After the test

	Activity
☐	Discuss your results with your teacher
☐	Identify your strengths and any potential areas to revise
☐	Consider these results together with other evidence of your progress
☐	Review learning goals to ensure they are appropriate

Calculator skills

NOTE: The following advice is based on common scientific calculators in use throughout Australian schools.

Basic keys

Key	Use
+ , − , x , ÷	Basic operations
=	Equals sign, gives the answer
·	Decimal point
DEL	Deletes previous entry
ANS	Retrieves previous answer
◄ ▲ ▼ ►	Allows us to move around the screen
MODE or SHIFT or 2ndF	Access other operations

Key	Use
(–) or +/–	Enters negative numbers
▭/▭ or a b/c	Enters fractions
( )	Enters parentheses
x^2	Squares a number
√	Finds the square root of a number
x^3	Cubes a number
∛	Finds the cube root of a number

Basic operations and using brackets

Question	Calculator steps	Answer
246 + 107 =	246 + 107 =	353
462 – 289 =	462 − 289 =	173
6 × 14 =	6 X 14 =	84
1463 ÷ 7 =	1463 ÷ 7 =	209
8 × (15 + 12) =	8 X (15 + 12) =	216
[(72-6) ÷ (25+8)] x 24 =	[(72 − 6) ÷ (25 + 8)] X 24 =	48
Find the average of 7.1, 3.6, 8 and 4.5	(7.1 + 3.6 + 8 + 4.5) ÷ 4 =	5.8

Integers

Question	Calculator steps	Answer
-6 + 15 =	+/– 6 + 15 =	9
-16 × 12 =	+/– 16 X 12 =	-192
8 – (-12) =	8 − +/– 12 =	20
-11 – 7 × -3 =	+/– 11 − 7 X +/– 3 =	10

9780170462204

Decimals

Question	Calculator steps	Answer
17.622 + 5.4 - 8.39 =	17 [·] 622 [+] 5 [·] 4 [−] 8 [·] 39 [=]	14.632
23.1 × 0.82 =	23 [·] 1 [×] 0 [·] 82 [=]	189.42
1.72 ÷ 0.8 =	1 [·] 72 [÷] 0 [·] 8 [=]	0.92
$7.60 × 9 =	7 [·] 6 [×] 9 [=]	$68.40
0.9 of 460 m =	0 [·] 9 [×] 460 [=]	414 m
55% × $186 =	0 [·] 55 [×] 186 [=]	$102.30
216% of 35 =	2 [·] 16 [×] 35 [=]	75.6

Powers and roots

Question	Calculator steps	Answer
$15^2 =$	15 [x^2] [=] or 15 [×] 15 [=]	225
$9^3 =$	9 [x^3] [=] or 9 [×] 9 [×] 9 [=]	729
$\sqrt{289} - \sqrt[3]{343} =$	[$\sqrt{\ }$] 289 [−] [$\sqrt[3]{\ }$] 343 [=]	10

Operations with fractions

Question	Calculator steps	Answer
Simplify: $\dfrac{56}{84}$	56 [▯] 84 [=]	$\dfrac{2}{3}$
Convert $2\frac{3}{8}$ to an improper fraction	2 [▯] 3 [▯] 8 [=] [SHIFT] [▯]	$\dfrac{19}{8}$
Convert $\dfrac{11}{5}$ to a mixed numeral	11 [▯] 5 [=]	$2\frac{1}{5}$
$3\frac{2}{5} + 4\frac{3}{10} =$	3 [▯] 2 [▯] 5+4 [▯] 3 [▯] 10 [=]	$7\frac{1}{10}$
$\dfrac{2}{5}$ of 250m =	2 [▯] 5 [×] 250 [=]	100 m

Test 1: Non-calculator

Instructions

- A correct answer scores 1 mark, and an incorrect answer scores 0.
- Marks are not deducted for incorrect answers.
- No marks are given if more than one answer alternative is shaded.
- Choose the alternative which most correctly answers the question and shade in the box next to it.

QUESTION 1

What time is the same as the time shown on this digital clock?

SHADE ONE BOX

15:45

☐ 3.45 a.m.　　　☐ 3.45 p.m.　　　☐ 5.45 a.m.　　　☐ 5.45 p.m.

QUESTION 2

What is the value of 241 + 362?

SHADE ONE BOX

☐ 503　　　☐ 593　　　☐ 603　　　☐ 613

QUESTION 3

Which fraction is equivalent to $\frac{3}{4}$?

SHADE ONE BOX

☐ $\frac{5}{8}$　　　☐ $\frac{9}{12}$　　　☐ $\frac{12}{18}$　　　☐ $\frac{20}{25}$

QUESTION 4

Peta draws this shape.

SHADE ONE BOX

A

She then rotates it 90° clockwise about A. After it has been rotated, what would it look like?

☐　　　☐　　　☐　　　☐ A

A　　　A　　　A

WRITE YOUR OWN ANSWER

Tom starts with a number. He adds 3 then multiplies it by 6. The answer is 42.

What number did Tom start with?

QUESTION 6

SHADE ONE BOX

What is $10 as a percentage of $50?

☐ 5%   ☐ 10%   ☐ 20%   ☐ 50%

QUESTION 7

WRITE YOUR OWN ANSWER

What is the missing number?

$4 \times \boxed{} = 5 \times 8$

QUESTION 8

SHADE ONE BOX

The column graph below shows the number of pets owned by students in a Year 7 class. Which of these statements is true?

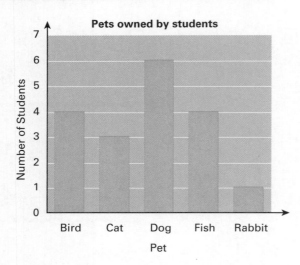

☐ The number of students who own cats is greater than the number who own birds.

☐ There are 20 students who own pets in this class.

☐ The number of students who own birds is smaller than the number who own rabbits.

☐ The number of students who own dogs is more than the number who own fish.

QUESTION 9

SHADE ONE BOX

Jeremy has a 1.5 L bottle of juice. He pours all the drink into 40 mL cups.
What is the best way to find the number of cups Jeremy can fill?

☐ 1.5 ÷ 100 × 40   ☐ 1.5 ÷ 1000 × 40   ☐ 1.5 × 100 ÷ 40   ☐ 1.5 × 1000 ÷ 40

QUESTION 10

SHADE ONE BOX

Which number is smaller than 8.7?

☐ 8.77   ☐ 8.07   ☐ 8.777   ☐ 8.707

QUESTION 11

SHADE ONE BOX

Hayden has a bag with 5 black balls, 2 red balls and 6 blue balls in it.
If he takes one coloured ball out of his bag, which word(s) describes the chance that it is red?

☐ certain

☐ very likely

☐ even chance

☐ unlikely

QUESTION 12

SHADE ONE BOX

Minna is using this biscuit recipe:

Anzac Biscuit recipe
Makes 20 biscuits

1 cup plain flour	$\frac{1}{2}$ cup butter
1 cup rolled oats	2 tbs golden syrup
1 cup desiccated coconut	1 tsp bicarbonate of soda
$\frac{1}{2}$ cup brown sugar	2 tbs boiling water

How many cups of butter are needed to make 100 Anzac biscuits?

☐ $\frac{1}{2}$ ☐ $2\frac{1}{2}$ ☐ 5   ☐ 10

QUESTION 13

SHADE ONE BOX

What is the best estimate of the size of this angle?

☐ 30°   ☐ 45°   ☐ 60°   ☐ 20°

QUESTION 14

SHADE ONE BOX

Which diagram shows the net of a cube?

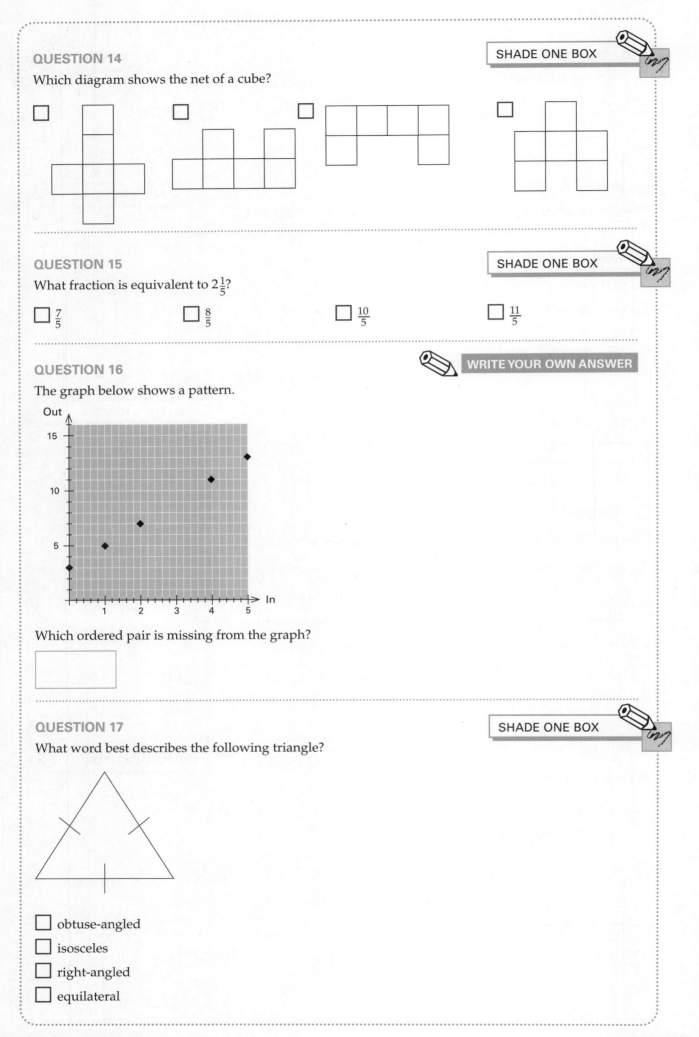

QUESTION 15

SHADE ONE BOX

What fraction is equivalent to $2\frac{1}{5}$?

☐ $\frac{7}{5}$ ☐ $\frac{8}{5}$ ☐ $\frac{10}{5}$ ☐ $\frac{11}{5}$

QUESTION 16

WRITE YOUR OWN ANSWER

The graph below shows a pattern.

Which ordered pair is missing from the graph?

QUESTION 17

SHADE ONE BOX

What word best describes the following triangle?

☐ obtuse-angled
☐ isosceles
☐ right-angled
☐ equilateral

QUESTION 18

Which number is three thousand, eight hundred and seventy?

SHADE ONE BOX

☐ 3870   ☐ 3807   ☐ 3087   ☐ 3078

QUESTION 19

An athlete walks at a rate of 3 m/s. At this rate, how far can the athlete walk in one minute?

SHADE ONE BOX

☐ 60 m   ☐ 180 m   ☐ 300 m   ☐ 360 m

QUESTION 20

This jug contains some cordial. If Chris takes 600 mL from the jug, how much cordial is left?

WRITE YOUR OWN ANSWER

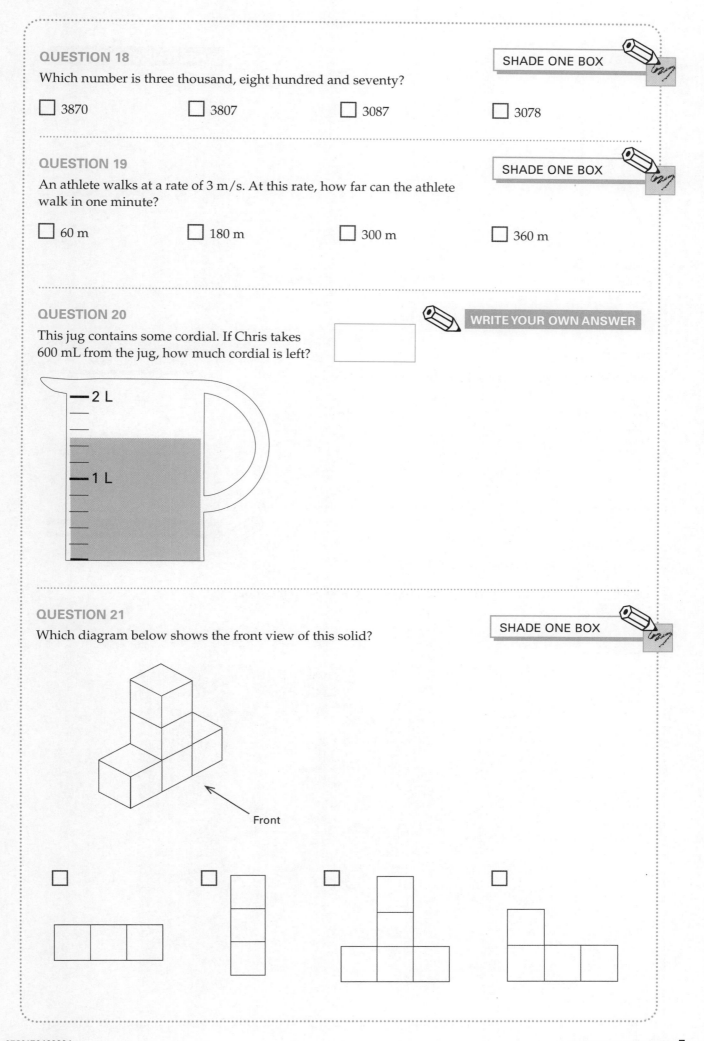

—2 L

—1 L

QUESTION 21

Which diagram below shows the front view of this solid?

SHADE ONE BOX

Front

☐   ☐   ☐   ☐

QUESTION 22

WRITE YOUR OWN ANSWER

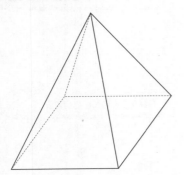

How many edges does this solid have?

QUESTION 23

SHADE ONE BOX

Which diagram below shows the radius of a circle?

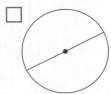 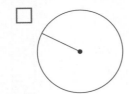 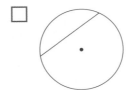 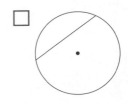

WRITE YOUR OWN ANSWER

QUESTION 24

The map below has a scale of 1 cm for every 3 km.

Use the map to find the actual distance from Anna's house to the shop.

km

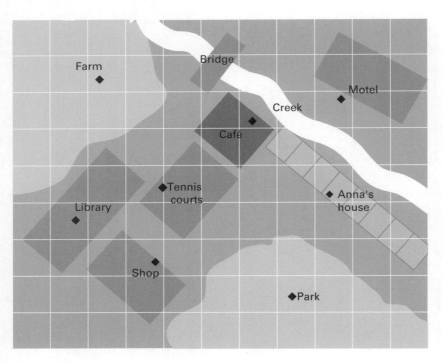

QUESTION 25

SHADE ONE BOX

Students at a school were surveyed to find out their favourite sport.
The results are shown in the pie chart below.

Students' favourite sports

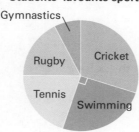

Which of the following statements is correct?

☐ One-quarter of the students preferred swimming.

☐ Tennis was more popular than cricket.

☐ One-third of the students preferred gymnastics.

☐ Cricket was less popular than rugby.

QUESTION 26

SHADE ONE BOX

Which spinner is most likely to spin red?

☐ 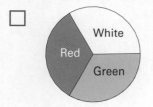   ☐ 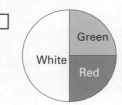   ☐ 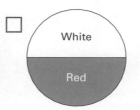   ☐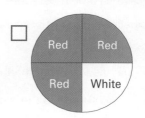

QUESTION 27

WRITE YOUR OWN ANSWER

For the set of quiz scores below, which is the mode?

8, 7, 7, 5, 4, 5, 6, 7, 8, 5, 9, 7

QUESTION 28

SHADE ONE BOX

The data in the table below compares the age of three children with their height.

Age	2	5	10
Height (cm)	95	115	148

Which graph best represents the data in the table?

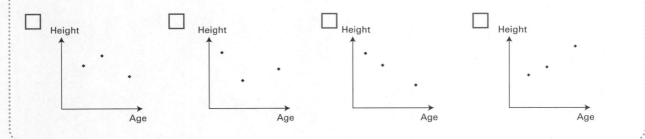

QUESTION 29

SHADE ONE BOX

60% of the fish in Andrew's fish tank are orange. If Andrew has 40 fish in the tank, how many fish are orange?

☐ 16   ☐ 24   ☐ 27   ☐ 30

QUESTION 30

SHADE ONE BOX

Which of the following expressions is equivalent to $4 \times 9 - 8 + 2$?

☐ $7 + 3 \times 6 + 5$ ☐ $18 \div 6 + 3$ ☐ $5 \times 4 + 3 \times 2$ ☐ $3^2 + 5^2$

QUESTION 31

SHADE ONE BOX

The first three terms in a sequence are 1, 4, 9, …, …. The next two terms will be:

☐ 11, 15   ☐ 13, 25   ☐ 15, 21   ☐ 16, 25

QUESTION 32

WRITE YOUR OWN ANSWER

Olivia, Phoebe and Georgia are sisters. They decide to put their pocket money together to buy their father a present for his birthday.

- Olivia has the most money.
- Phoebe has half of the amount of money that Georgia has.
- Georgia has $22.
- Olivia has $15 more than Phoebe.

How much money do the sisters have to spend on the present?

Test 2: Calculator allowed

Instructions

- A correct answer scores 1 mark, and an incorrect answer scores 0.
- Marks are not deducted for incorrect answers.
- No marks are given if more than one answer alternative is shaded.
- Choose the alternative which most correctly answers the question and shade in the box next to it.

QUESTION 1

SHADE ONE BOX

Which number is 6 thousandths?

☐ 0.06   ☐ 0.006   ☐ 0.6   ☐ 6000

QUESTION 2

SHADE ONE BOX

What is the value of 762 – 654 equivalent to?

☐ 102   ☐ 108   ☐ 112   ☐ 118

QUESTION 3

SHADE ONE BOX

The area of a rectangle is 306 cm². If its length is 18 cm, what is its width?

Not to scale

18 cm

☐ 3708 cm
☐ 17 cm
☐ 648 cm
☐ 19 cm

QUESTION 4

This graph shows which students play hockey and tennis at a school.

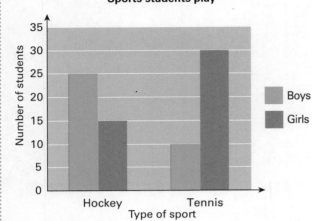

Use the information in the graph to complete this table.

	Hockey	Tennis
Boys		10
Girls	15	

QUESTION 5

David walked for 30 seconds. He walked at a speed of 5 m per second.

How far did David walk?

☐ 6 m   ☐ 35 m   ☐ 150 m   ☐ 200 m

QUESTION 6

What is $\frac{7}{10} - \frac{1}{10}$ equivalent to?

☐ $\frac{3}{4}$ ☐ $\frac{3}{5}$ ☐ $\frac{1}{2}$ ☐ $\frac{2}{5}$

QUESTION 7

$\triangle + \triangle \times \square = 48$

$\triangle + \square = 15$

For the number sentences shown above, what are the values of $\triangle$ and $\square$?

☐ $\triangle = 4$ and $\square = 11$

☐ $\triangle = 6$ and $\square = 9$

☐ $\triangle = 8$ and $\square = 5$

☐ $\triangle = 2$ and $\square = 13$

QUESTION 8

How many hours and minutes are there between 9.36 a.m. and 4.21 p.m. on the same day?

☐ 6 h 45 min   ☐ 6 h 57 min   ☐ 7 h 45 min   ☐ 7 h 57 min

QUESTION 9

The number of trees planted in the suburb of Lyneville over five years is shown below:

Trees Planted in Lyneville

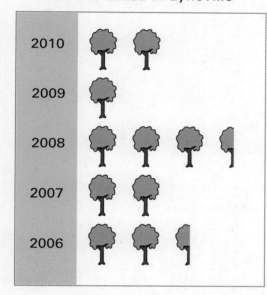

What is the total number of trees planted over the five years?

☐ 600

☐ 550

☐ 500

☐ 450

QUESTION 10

The ratio of rainy days to sunny days over 140 consecutive days is 3 to 7.

How many rainy days were there?

☐ 42   ☐ 60   ☐ 98   ☐ 112

QUESTION 11

This is a map of an amusement park. The main attractions are shown on the map below.

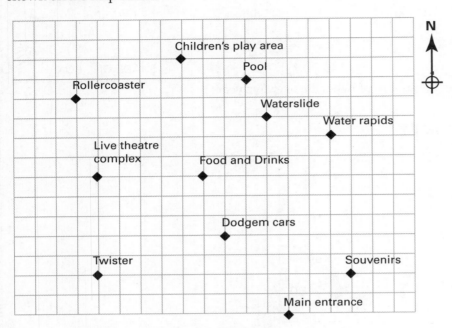

Joshua is at the food and drinks stand. What attraction is positioned north-east of Joshua?

QUESTION 12

SHADE ONE BOX

What is the same as 1 L and 60 mL?

☐ 160 L   ☐ 1.06 L   ☐ 1006 mL   ☐ 1.6 mL

QUESTION 13

SHADE ONE BOX

Which two objects have exactly the same number of edges?

A   B   C   D

☐ A and D

☐ A and C

☐ B and C

☐ C and D

QUESTION 14

WRITE YOUR OWN ANSWER

The table below shows the temperature in Hazelfields over five days.

Temperature in Hazelfields	
Day	**Temperature (°C)**
Monday	23
Tuesday	25
Wednesday	21
Thursday	28
Friday	33

What is the average (mean) temperature for these five days?

Average =

QUESTION 15

SHADE ONE BOX

In the shape below, what is the total length of the two sides with dotted lines?

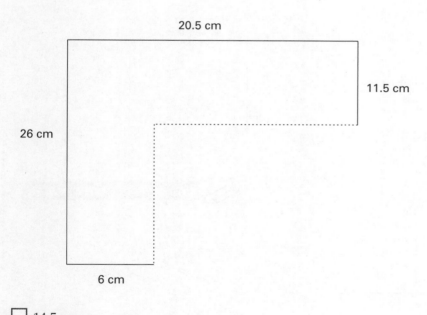

- [ ] 14.5 cm
- [ ] 17.5 cm
- [ ] 29 cm
- [ ] 31 cm

QUESTION 16

SHADE ONE BOX

A movie theatre can hold 400 people. At a movie screening, 236 people watched the movie. The percentage of movie theatre seats that were occupied during this movie screening is closest to:

- [ ] 40%
- [ ] 50%
- [ ] 60%
- [ ] 70%

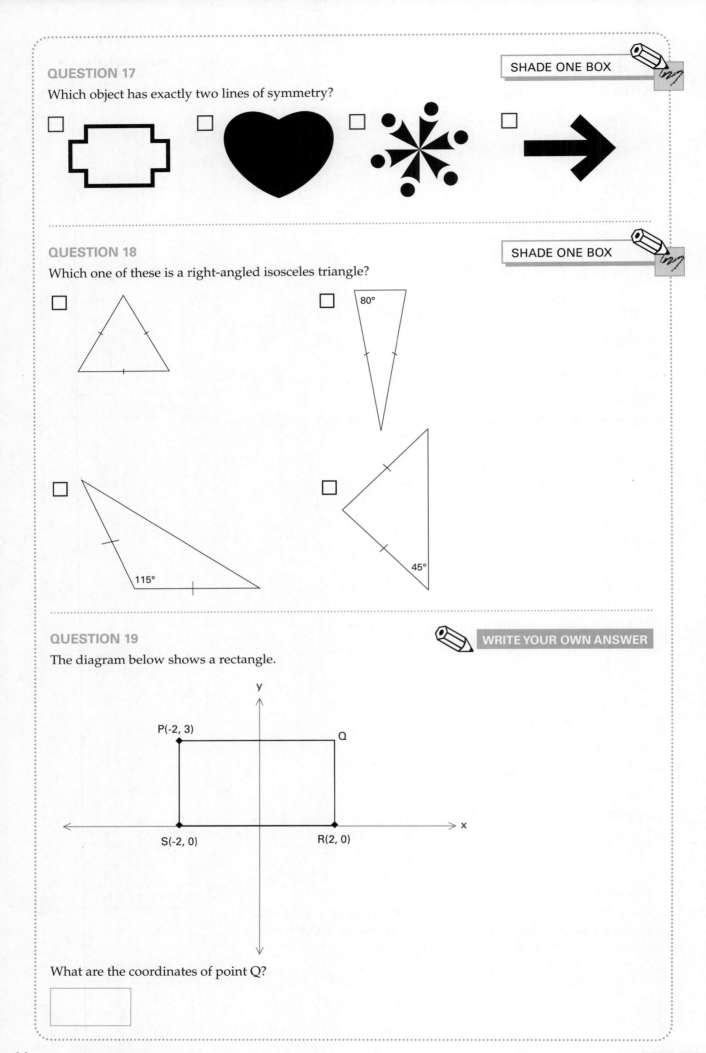

QUESTION 17

SHADE ONE BOX

Which object has exactly two lines of symmetry?

☐ ☐ ☐ ☐

QUESTION 18

SHADE ONE BOX

Which one of these is a right-angled isosceles triangle?

☐ ☐ 80°

☐ 115° ☐ 45°

QUESTION 19

WRITE YOUR OWN ANSWER

The diagram below shows a rectangle.

P(-2, 3)   Q

S(-2, 0)   R(2, 0)

What are the coordinates of point Q?

9780170462204

QUESTION 20

WRITE YOUR OWN ANSWER

Megan bought these items from the supermarket:

- 1 kg grapes, at $7.50 per kilogram
- 2 bottles of juice, $1.43 each
- 1 magazine, $6.80
- 3 loaves of bread, $2.90 each
- 2.5 kg of sausages, at $6.70 per kilogram.

How much change will Megan receive from a $50 note? Answer to nearest five cents.

$ [   ]

QUESTION 21

SHADE ONE BOX

Emma is making this pattern out of sticks.

  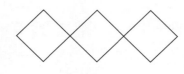

| Shape 1 | Shape 2 | Shape 3 |

The table below shows the number of sticks she needs for each shape in the pattern.

Shape	1	2	3	4
Number of Sticks	4	8	12	?

How many sticks are needed to make shape 4?

- ☐ 15
- ☐ 16
- ☐ 20
- ☐ 24

QUESTION 22

SHADE ONE BOX

What is the value of $\frac{1}{4} + \frac{3}{5}$?

- ☐ $\frac{1}{5}$
- ☐ $\frac{4}{9}$
- ☐ $\frac{7}{10}$
- ☐ $\frac{17}{20}$

QUESTION 23

Harry spins these two arrows. He adds the numbers in the sections where the arrows stop and gets a sum of 5.

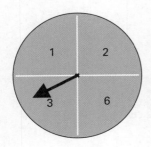 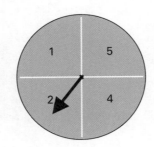

Harry then spins the arrows again. In how many different ways can he get a sum of 7?

☐ 1   ☐ 2   ☐ 3   ☐ 4

QUESTION 24

Stephen's water bill was $160 last month. This month it is $120. What percentage decrease is this?

☐ 20%   ☐ 25%   ☐ 40%   ☐ 75%

QUESTION 25

WRITE YOUR OWN ANSWER

Here is a map of Camper's Island.

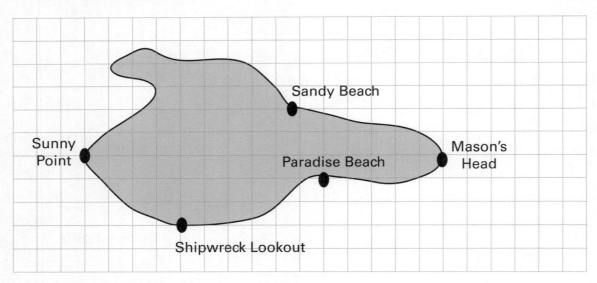

If 1 unit is equivalent to 1.5 km, find the shortest distance between Sunny Point and Mason's Head.

☐ km

QUESTION 26

SHADE ONE BOX

I start with a number and add 6. I then multiply the answer by 3 and subtract 11. The answer is 61. What is the number I started with?

☐ 25   ☐ 24   ☐ 18   ☐ 12

QUESTION 27

SHADE ONE BOX

Which two shapes have the same perimeter?

A   B   C   D

8 cm   8 cm   12 cm   8 cm   8 cm

☐ A and C

☐ B and C

☐ B and D

☐ A and D

QUESTION 28

SHADE ONE BOX

Marcus counted 60 vehicles that crossed a bridge. This table shows the percentage of each type of vehicle that crossed the bridge.

Type of vehicle	Percentage
Car	45%
Bus	35%
Truck	15%
Motorbike	5%

How many buses crossed the bridge?

☐ 18

☐ 21

☐ 27

☐ 35

QUESTION 29

SHADE ONE BOX

A school canteen made 2 L of hot chocolate to sell at recess. Each cup of hot chocolate was 250 mL. If Vanessa and her six friends each bought a cup at recess, how much hot chocolate was left?

☐ 0.25 L   ☐ 0.5 L   ☐ 1.5 L   ☐ 1.75 L

QUESTION 30

The graph below shows the height of a hot-air balloon after it leaves the ground.

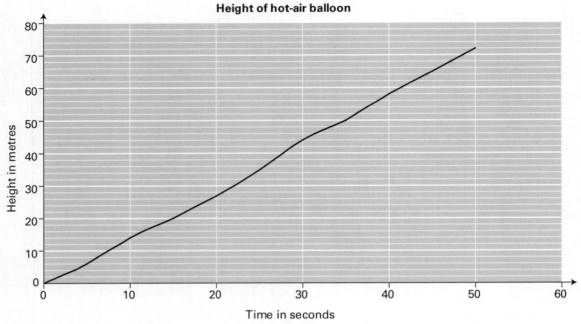

From the graph, find the height of the hot-air balloon after 30 seconds.

| | m

QUESTION 31

WRITE YOUR OWN ANSWER

Two friends, Lindsay and Harun, ordered a large pizza. A large pizza has 12 slices. If Lindsay ate one-third of the pizza and Harun ate one-quarter, how many slices of pizza did the boys eat altogether?

| | slices

QUESTION 32

SHADE ONE BOX

A farmer counted the total number of heads and legs of his sheep and chickens. There were 20 heads and 64 legs altogether. How many chickens did the farmer have?

☐ 8   ☐ 10   ☐ 11   ☐ 12

Test 3: Non-calculator

Instructions

- A correct answer scores 1 mark, and an incorrect answer scores 0.
- Marks are not deducted for incorrect answers.
- No marks are given if more than one answer alternative is shaded.
- Choose the alternative which most correctly answers the question and shade in the box next to it.

QUESTION 1

SHADE ONE BOX

What time is it $2\frac{1}{2}$ hours before 10.20 p.m.?

☐ 7:50 p.m.　　　☐ 12:50 a.m.　　　☐ 7:30 p.m.　　　☐ 12:30 a.m.

QUESTION 2

SHADE ONE BOX

How many of these numbers are multiples of 7?

59　　14　　84　　72　　49

42　　56　　48　　21　　70　　62

☐ 5　　　　☐ 6　　　　☐ 7　　　　☐ 8

QUESTION 3

SHADE ONE BOX

This plan shows the layout of shops in a shopping centre.

A1		A2	H2		F4
B1		B2	G2		E4
C1		C2	F2		D4
D1		D2	E2		C4

| E1 | | A3 | B3 | C3 | D3 | | B4 |
| F1 | | | | | | | A4 |

Tina enters the shopping centre at the arrow. She takes the second turn on her left and goes into the second shop on her right. Which shop did Tina enter?

☐ C2　　　　☐ F2　　　　☐ A4　　　　☐ D4

QUESTION 4

SHADE ONE BOX

A triangular prism and a rectangular prism have been glued together.

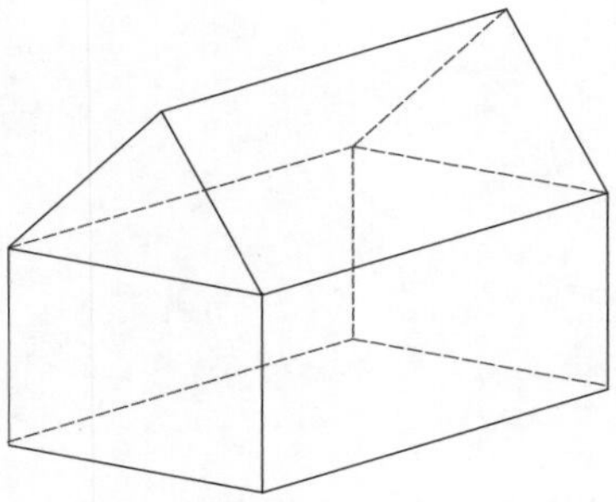

How many faces does the new object have?

☐ 4 ☐ 7 ☐ 8 ☐ 11

QUESTION 5

WRITE YOUR OWN ANSWER

Sam's birthday is 8 April and Chloe's birthday is 11 July. How many days are there from Sam's birthday to Chloe's?

☐ days

QUESTION 6

SHADE ONE BOX

$4 \times 1000 + 9 \times 100 + 6 =?$

☐ 496 ☐ 4096 ☐ 4906 ☐ 4960

QUESTION 7

SHADE ONE BOX

The object shown below was made using identical cubes.

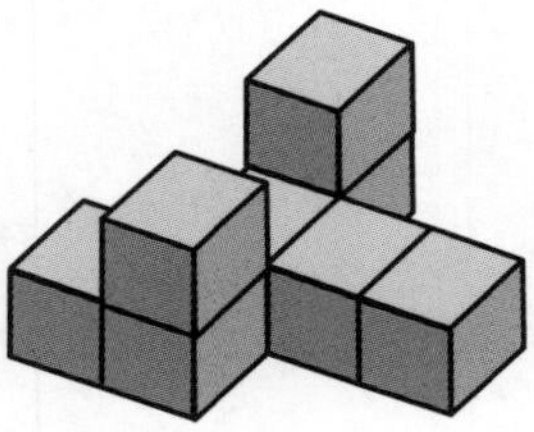

Which diagram below shows the top view of this solid?

☐ ☐ ☐ ☐

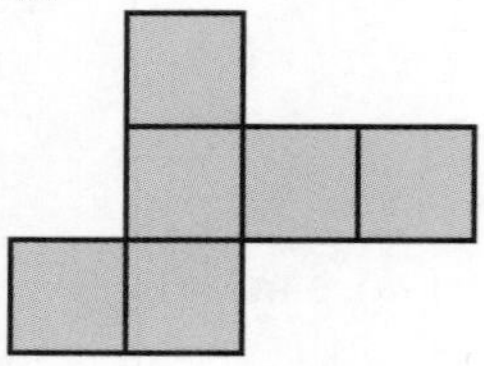

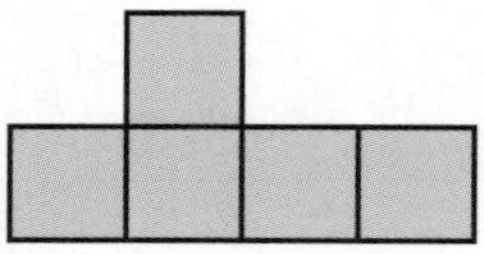

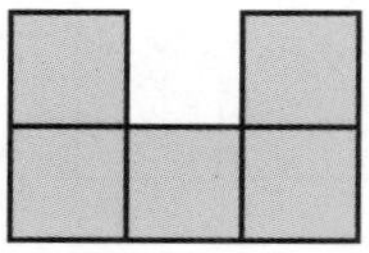

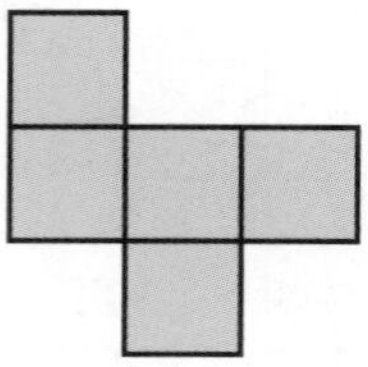

Andrew takes one ball out of his bag without looking. It is very unlikely, but not impossible, that he will get a white ball. Which bag is Andrew's?

QUESTION 9

Convert 0.391 kg to grams.

☐ 3.91 g ☐ 39.1 g ☐ 391 g ☐ 3910 g

QUESTION 10

Look at these dogs.

What fraction of these dogs are not black?

☐ $\frac{3}{5}$ ☐ $\frac{5}{8}$ ☐ $\frac{2}{3}$ ☐ $\frac{3}{4}$

QUESTION 11

Zoe is facing south-west and turns 90° in a clockwise direction. Which direction is she now facing?

☐ south-east ☐ north-west ☐ east ☐ west

QUESTION 12

The table shows the results of a survey on Internet usage.

	Monthly Internet usage		
Age	**10 h or less**	**more than 10 h but less than 20 h**	**20 h or more**
Under 20 years	8	26	23
20–40 years	15	19	11
40–60 years	22	15	14
over 60 years	10	4	0

In total, how many people aged 20–40 years used the Internet each month for more than 10 hours?

☐ people

QUESTION 13

SHADE ONE BOX

A number line is shown below.

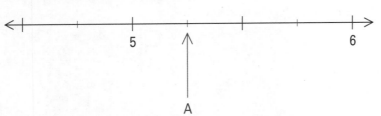

What number is represented on the number line by A?

☐ $5\frac{1}{4}$ ☐ $5\frac{3}{8}$ ☐ $5\frac{1}{2}$ ☐ $5\frac{3}{4}$

QUESTION 14

WRITE YOUR OWN ANSWER

The maximum daily temperatures (°C) in Prague and London over one week in January are shown below.

	Sunday	Monday	Tuesday	Wednesday	Thursday	Friday	Saturday
Prague	-6	-4	-5	-3	-1	1	0
London	3	4	7	2	-1	-2	3

Which day shows the greatest difference in temperature between the two cities?

QUESTION 15

WRITE YOUR OWN ANSWER

Rashid is using this scone recipe:

Scone recipe

Makes 20 scones   60 g butter

2 cups flour   1 cup milk

$\frac{1}{4}$ tsp salt

How many cups of flour are needed to make 70 scones?

☐ cups

QUESTION 16

What is 1256 ÷ 8 equivalent to?

SHADE ONE BOX

☐ 107 ☐ 132 ☐ 146 ☐ 157

QUESTION 17

The rule for the table of values given below is y = 2x + 3.

SHADE ONE BOX

x	1	3	5	7	9
y	5	9	?	17	21

What value is missing?

☐ 12 ☐ 13 ☐ 14 ☐ 15

9780170462204

QUESTION 18

SHADE ONE BOX

Which one of these expressions has a value of 4?

☐ $18 - 10 \div 2$

☐ $\frac{3}{12}$

☐ $4^3 \div 4^2$

☐ $8 \times 4 \div 2 + 6$

QUESTION 19

SHADE ONE BOX

Gemma draws this shape:

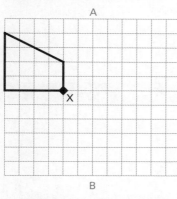

She reflects it in the line AB and then rotates it 90° clockwise about point X.

What would it look like after it has been reflected and rotated?

☐ 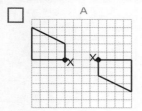 ☐ 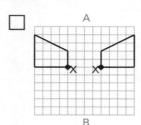 ☐ 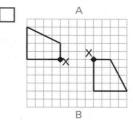 ☐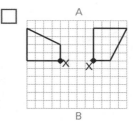

QUESTION 20

SHADE ONE BOX

Which decimal is closest to 0.74?

☐ 0.75   ☐ 0.71   ☐ 0.8   ☐ 0.7

QUESTION 21

WRITE YOUR OWN ANSWER

What is the ordered pair that represents point B?

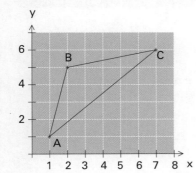

QUESTION 22

 WRITE YOUR OWN ANSWER

Jake wants to make a solid cube that is three small cubes long, wide and high.

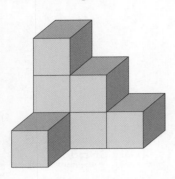

How many **more** cubes does Jake need?

	cubes

QUESTION 23

 WRITE YOUR OWN ANSWER

I start with the number 7. I then add 2 and multiply the answer by 8. Then I subtract 15. What is the answer?

QUESTION 24

SHADE ONE BOX

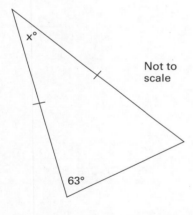

Not to scale

What is the value of × in the diagram above?

☐ 54   ☐ 63   ☐ 117   ☐ 126

QUESTION 25

 WRITE YOUR OWN ANSWER

Peter places 6 books in each of 9 boxes.

He has 4 books leftover.

Peter wants to place 8 books in each box. How many more books does he need?

	books

QUESTION 26

SHADE ONE BOX

Half a cake has been eaten. The remainder is to be shared equally between three children. Which diagram shows the amount of cake each child will receive?

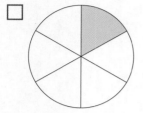 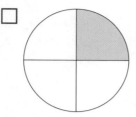 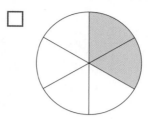 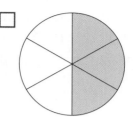

QUESTION 27

WRITE YOUR OWN ANSWER

On a busy night in a restaurant, the chef counted the meals made. The results are shown in the column graph below.

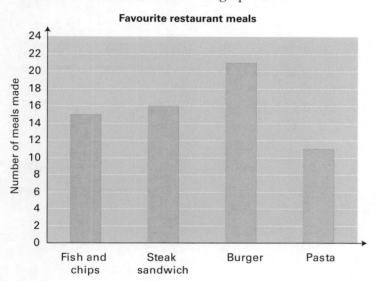

How many meals were made that night?

| | meals

QUESTION 28

SHADE ONE BOX

The diagram below shows a rectangle and a triangle joined together.

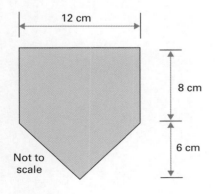

12 cm

8 cm

6 cm

Not to scale

What is the total area of this shape?

☐ 168 cm²   ☐ 132 cm²   ☐ 120 cm²   ☐ 84 cm²

QUESTION 29

This tile pattern is made up of squares and rectangles. Each rectangle has a length of 10 cm and a width of 5 cm.

SHADE ONE BOX

What is the total area of one tile?

☐ 200 cm²   ☐ 225 cm²   ☐ 250 cm²   ☐ 275 cm²

QUESTION 30

SHADE ONE BOX

On a holiday adventure camp, the ratio of children to cabins is 4 to 1.
If there are eight cabins, how many children are there?

☐ 32   ☐ 24   ☐ 16   ☐ 12

QUESTION 31

SHADE ONE BOX

A cyclist can ride one lap of a bike track in 40 seconds. How many laps will the cyclist have completed after riding for four minutes at the same speed?

☐ 6   ☐ 8   ☐ 10   ☐ 15

QUESTION 32

WRITE YOUR OWN ANSWER

In a test, three marks are awarded for every correct answer and one mark is taken away for each wrong answer. If there are 20 questions in the test and David is awarded 36 marks, how many questions did he answer correctly?

[_____] questions correct

Test 4: Calculator allowed

Instructions

- A correct answer scores 1 mark, and an incorrect answer scores 0.
- Marks are not deducted for incorrect answers.
- No marks are given if more than one answer alternative is shaded.
- Choose the alternative which most correctly answers the question and shade in the box next to it.

QUESTION 1

Which of the following shapes is a pentagon?

SHADE ONE BOX

☐   ☐   ☐   ☐

QUESTION 2

Which of these numbers is the smallest?

SHADE ONE BOX

☐ 3.101   ☐ 3.1101   ☐ 3.10001   ☐ 3.11001

QUESTION 3

Sabah bought 7 bottles of milk at $2.35 each and 4 cans of tomatoes at $2.99 each. How much change did she get from $50? Assume you can still get one- and two-cent coins.

SHADE ONE BOX

☐ $28.41   ☐ $27.76   ☐ $21.59   ☐ $38.29

QUESTION 4

WRITE YOUR OWN ANSWER

A square has a side length of 4.5 cm. What is the area of the square in square metres?

☐ square metres

QUESTION 5

Vito has 6 pairs of black socks, 2 pairs of white socks and 3 pairs of red socks in his drawer. He chooses a pair of socks without looking. Which of the following statements below is correct?

SHADE ONE BOX

☐ Vito has an equal chance of choosing a red or a black pair of socks.

☐ Vito cannot choose a pair of white socks.

☐ Vito is likely to choose a black pair of socks.

☐ Vito is certain to choose a black pair of socks.

QUESTION 6

The city of Kingsbury has the following information about recycling in its annual report. Which of the following statements is true?

Kingsbury trash

☐ Yard trimmings and food scraps account for more than one quarter of the recycling.

☐ Plastics and yard trimmings account for more than $\frac{1}{3}$ of the recycling.

☐ Paper and 'other' account for more than half the recycling.

☐ Metals, food scraps and glass account for less than $\frac{1}{5}$ of the recycling.

QUESTION 7

Which of the points below has the coordinates (2, 4)?

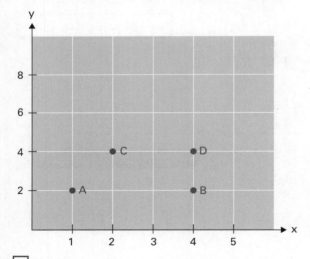

☐ A
☐ B
☐ C
☐ D

QUESTION 8

WRITE YOUR OWN ANSWER

George bought a plasma TV for $2599. It was originally advertised for $3050. What percentage discount of the original price did he receive? Write your answer to the nearest whole number.

☐☐☐☐☐ %

QUESTION 9

SHADE ONE BOX

Charlotte catches the 4.03 p.m. train from Reservoir and, after transferring to a tram, arrives at her grandma's house in St Kilda at 5.17 p.m. How long did her journey take?

☐ 14 min   ☐ 1 h 15 min   ☐ 1 h 14 min   ☐ 15 min

QUESTION 10

SHADE ONE BOX

A camera took a picture of some geometric solids as shown below. Notice the position of the camera.

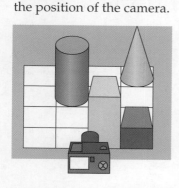

Which of the following images represents the arrangement of the solids taken from another viewpoint? Notice the camera has moved.

☐ 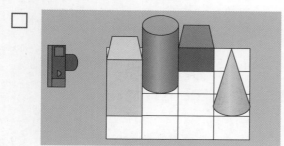   ☐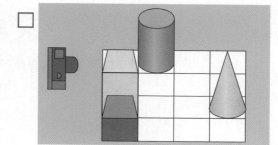

☐ 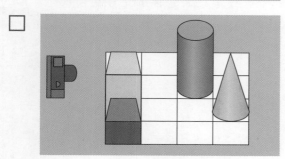   ☐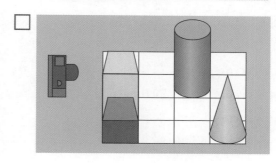

QUESTION 11

SHADE ONE BOX

Here are the first four terms of a dots-and-triangles pattern.

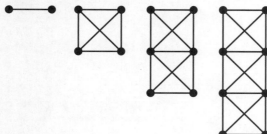

How many small triangles would there be in a pattern made with 12 dots?

☐ 20   ☐ 16   ☐ 24   ☐ 18

QUESTION 12

SHADE ONE BOX

What is the highest common factor of 24 and 30?

☐ 3   ☐ 12   ☐ 6   ☐ 8

QUESTION 13

SHADE ONE BOX

When 7.81 km is added to 6300 m, what is the result?

☐ 8.43 km   ☐ 7.863 km   ☐ 14.11 km   ☐ 13.83 km

QUESTION 14

SHADE ONE BOX

Below is the bus timetable from Melbourne to La Trobe University.

Bus stop	Time at bus stop (a.m.)							
Melbourne (Bourke/Queen)	6.40	6.58	7.10	7.35	7.45	7.54	8.05	8.16
Melbourne (Spring/Flinders)	6.46	7.04	7.16	7.41	7.52	8.01	8.12	8.23
Carlton	6.52	7.11	7.23	7.48	7.59	8.08	8.19	8.30
Fitzroy North	7.06	7.25	7.40	8.03	8.14	8.24	8.34	8.45
Preston	7.17	7.37	7.50	8.14	8.25	8.36	8.45	8.56
Reservoir	7.29	7.49	8.01	8.27	8.37	8.49	8.56	9.07
La Trobe University	7.36	7.57	8.09	8.35	8.46	8.57	9.05	9.15

Lynise must reach La Trobe no later than 9.00 a.m. What is the latest bus he can catch from Carlton?

☐ 8.19 a.m.   ☐ 8.08 a.m.   ☐ 8.24 a.m.   ☐ 7.59 a.m.

QUESTION 15

SHADE ONE BOX

A fair coin is tossed 100 times. Which of the following events is least likely to occur?

☐ Heads occurs exactly 22 times.

☐ Heads occurs exactly 70 times.

☐ Heads occurs exactly 54 times.

☐ Heads occurs exactly 89 times.

QUESTION 16

WRITE YOUR OWN ANSWER

Write this number in decimal form.

$6 \times 10 + 4 \times \frac{1}{10} + 9 \times \frac{1}{1000}$

QUESTION 17

WRITE YOUR OWN ANSWER

Work out the total cost of 3 m of fabric at $12.95 per metre, a reel of cotton for $2.45 and a zipper for $19.95.

$

QUESTION 18

SHADE ONE BOX

Which one of these is equal to 2^3?

☐ 2×3 ☐ 3×3 ☐ $2 \times 2 \times 2$ ☐ $2 + 2 + 2$

QUESTION 19

WRITE YOUR OWN ANSWER

Mannix sold small teddy bears to raise money for a children's
charity. Each teddy bear was sold for $5. The charity received 60c for each teddy bear sold.
If Mannix raised $90 for the charity, how many teddy bears did he sell?

☐ teddy bears

QUESTION 20

WRITE YOUR OWN ANSWER

A tap leaks at a rate of 30 mL every 20 seconds. How much water
is wasted in five minutes?

☐ mL

QUESTION 21

SHADE ONE BOX

Jason is an athlete in training for a triathlon. He divides up his weekly
training sessions as shown in the bar graph below.

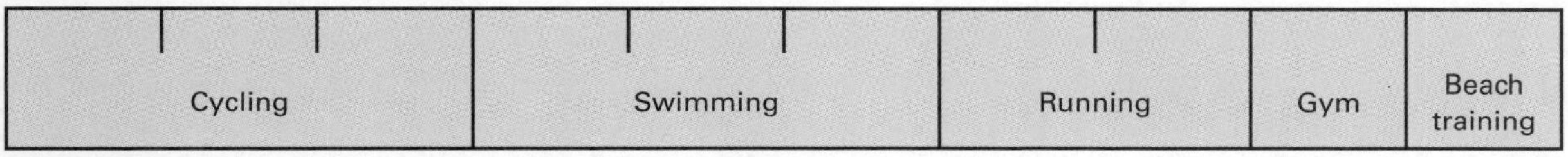

Cycling	Swimming	Running	Gym	Beach training

If Jason spends 18 hours each week running, how long does he spend cycling?

☐ 12 h   ☐ 15 h   ☐ 27 h   ☐ 30 h

QUESTION 22

WRITE YOUR OWN ANSWER

A kitchen floor is to be tiled. The area to be tiled is 3.5 m wide
by 4 m long. The tiles to be used are 20 cm × 35 cm. Each tile
costs $14.50. What is the total cost of the tiles?

QUESTION 23

SHADE ONE BOX

$1^2 + 2^3 + 3^4 = ?$

☐ 20   ☐ 36   ☐ 82   ☐ 90

QUESTION 24

SHADE ONE BOX

The volume of this rectangular prism is 192 cm³.

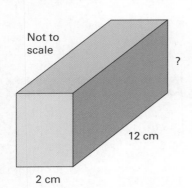

Not to scale

?

12 cm

2 cm

What is the length of this rectangular prism?

☐ 6 cm   ☐ 8 cm   ☐ 12 cm   ☐ 14 cm

QUESTION 25

SHADE ONE BOX

Look at the table below. The rule for this table is output = 3 × input + 5.

Input	Output
5	20
10	?
15	50
20	65

What is the missing value in the table?

☐ 45   ☐ 40   ☐ 35   ☐ 30

QUESTION 26

 WRITE YOUR OWN ANSWER

The table shows the amount of time Tanya spent studying over five days.

Day	Time
Monday	25 min
Tuesday	40 min
Wednesday	48 min
Thursday	1 h
Friday	27 min

What was the average time Tanya spent studying?

[ ] minutes

9780170462204

WRITE YOUR OWN ANSWER

QUESTION 27

This year, 15476 people visited the aquarium. This is 1532 more people than last year. How many people visited the aquarium last year?

☐ people

SHADE ONE BOX

QUESTION 28

Three students were asked to toss a coin 10 times. One of them thought it was a bit boring and decided to fake the results. Which of these results is most likely to be faked?

☐ T H H H T T H T H H

☐ T H T H H H T H T H T

☐ T T T T H H T H T H

☐ T H T H T H T H T H

SHADE ONE BOX

QUESTION 29

Which of the following drinks has the weakest mix of cordial?

☐ 120 mL water and 7 mL cordial

☐ 80 mL water and 3 mL cordial

☐ 100 mL water and 5 mL cordial

☐ 150 mL water and 6 mL cordial

SHADE ONE BOX

QUESTION 30

About what fraction of Jo's height is her head?

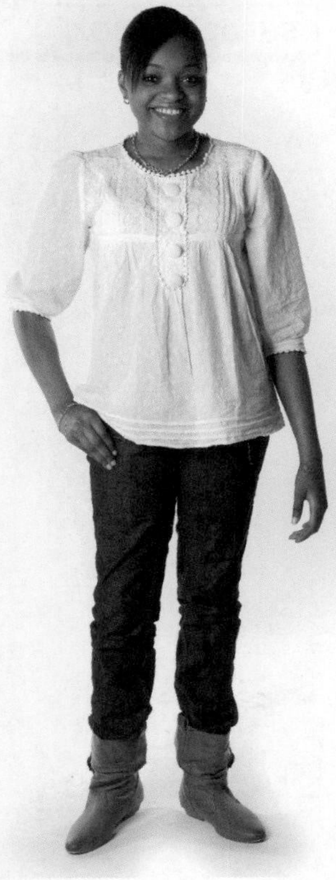

☐ one-eighth ☐ one-sixth ☐ one-tenth ☐ one-fifth

QUESTION 31

Which of the following is a graph of y = 2x + 2

□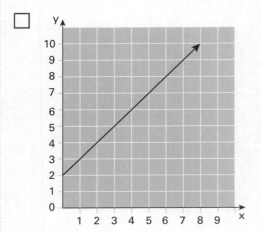

□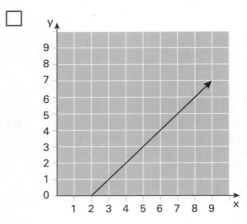

□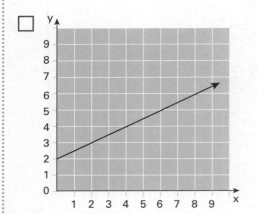

□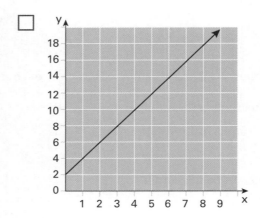

SHADE ONE BOX

QUESTION 32

What is the smallest angle and direction of turn to make P the image of
S in the diagram below?

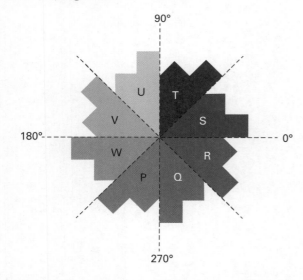

□ 135° anticlockwise   □ 135° clockwise   □ 225° anticlockwise   □ 225° clockwise

9780170462204

Test 5: Non-calculator

Instructions

- A correct answer scores 1 mark, and an incorrect answer scores 0.
- Marks are not deducted for incorrect answers.
- No marks are given if more than one answer alternative is shaded.
- Choose the alternative which most correctly answers the question and shade in the box next to it.

QUESTION 1

SHADE ONE BOX

If m = 7, what is the value of 2m?

☐ 14 ☐ 21 ☐ 27 ☐ 72

QUESTION 2

SHADE ONE BOX

What is another way of writing 5^3?

☐ 5 × 3
☐ 5 × 5 × 5
☐ 5 + 5 + 5
☐ 3 × 3 × 3 × 3 × 3

QUESTION 3

WRITE YOUR OWN ANSWER

There are 12 marbles in a bag. Four of the marbles are blue and
the rest are yellow. Casey chooses a marble from the bag without
looking. What is the chance of her choosing a **yellow** marble, written in simplest terms?

QUESTION 4

SHADE ONE BOX

What is the answer to 0.96 ÷ 0.4?

☐ 0.024 ☐ 0.24 ☐ 2.4 ☐ 24

QUESTION 5

SHADE ONE BOX

$\frac{1}{4} + \frac{7}{9} =$

☐ $\frac{7}{36}$ ☐ $\frac{8}{13}$ ☐ $\frac{29}{36}$ ☐ $1\frac{1}{36}$

QUESTION 6

The pie chart below shows the number of people living in each Australian state and territory.

Population of Australian states and territories

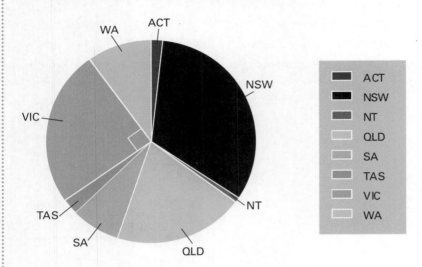

Which state's population represents approximately one quarter of Australia's total population?

☐ New South Wales

☐ Victoria

☐ Queensland

☐ Tasmania

QUESTION 7

Two glasses together contain 450 mL of milk. The first glass contains 70 mL less than the second glass. How much milk does the first glass contain?

☐ 190 mL ☐ 260 mL ☐ 330 mL ☐ 380 mL

QUESTION 8

How many hours and minutes is it between 13:12 Monday and 07:25 Tuesday?

☐ 17 h 13 min

☐ 17 h 37 min

☐ 18 h 13 min

☐ 18 h 37 min

What is the best way to estimate the total cost of these four objects?

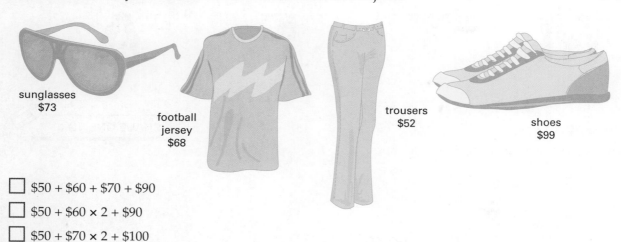

sunglasses
$73

football
jersey
$68

trousers
$52

shoes
$99

☐ $50 + $60 + $70 + $90

☐ $50 + $60 × 2 + $90

☐ $50 + $70 × 2 + $100

☐ $50 + $60 + $70 + $100

QUESTION 10

 WRITE YOUR OWN ANSWER

What is the shaded area of this diagram, in square units?

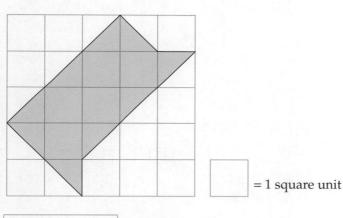

☐ = 1 square unit

┌──────────┐
│   │   square units
└──────────┘

QUESTION 11

A regular six-sided die is rolled once. What is the chance of getting a number less than 3?

☐ impossible   ☐ unlikely   ☐ likely   ☐ certain

QUESTION 12

SHADE ONE BOX

Which letter does not have a line of symmetry?

☐ **M**   ☐ **T**   ☐ **F**   ☐ **A**

QUESTION 13

SHADE ONE BOX

This object was made using identical cubes.

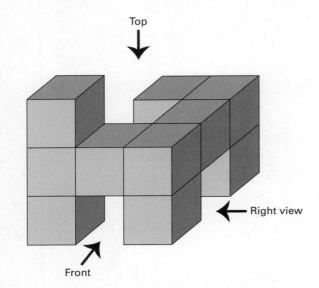

Which drawing shows the view from the top?

☐   ☐ 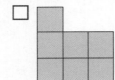   ☐ 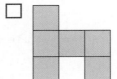   ☐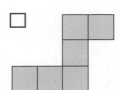

QUESTION 14

SHADE ONE BOX

What is the rule for the table of values shown below?

m	3	4	5	6
n	13	18	23	28

☐ $n = 5 \times m - 2$

☐ $n = m + 5$

☐ $n = 2 \times m + 5$

☐ $n = 5 \times (m - 2)$

9780170462204

What is the size of ∠ABC in the diagram?

[ ] degrees

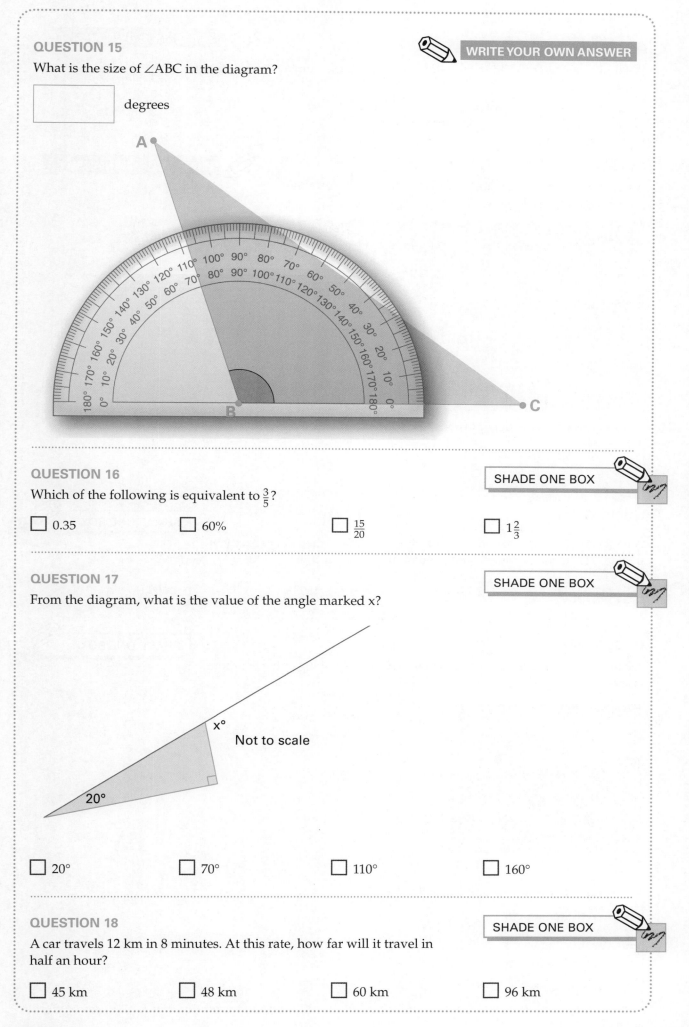

QUESTION 16

Which of the following is equivalent to $\frac{3}{5}$?

SHADE ONE BOX

☐ 0.35   ☐ 60%   ☐ $\frac{15}{20}$ ☐ $1\frac{2}{3}$

QUESTION 17

From the diagram, what is the value of the angle marked x?

SHADE ONE BOX

x°

Not to scale

20°

☐ 20°   ☐ 70°   ☐ 110°   ☐ 160°

QUESTION 18

SHADE ONE BOX

A car travels 12 km in 8 minutes. At this rate, how far will it travel in half an hour?

☐ 45 km   ☐ 48 km   ☐ 60 km   ☐ 96 km

WRITE YOUR OWN ANSWER

What number can be used in both boxes to make these number sentences equal?

$\boxed{} \times 6 + 2 = (\boxed{} - 2) \times 8$ $\boxed{}$

WRITE YOUR OWN ANSWER

A school held its annual sports carnival. This table shows the points scored by each house.

Field event	House				
	Red	Blue	Green	Yellow	Purple
Javelin	15	20	14	12	17
Shot put	12	15	16	20	19
High jump	16	20	11	14	15
Long jump	13	14	20	17	18
Discus	17	18	16	18	20

Which house won the field events at the sports carnival?

$\boxed{}$

SHADE ONE BOX

A tank contains 360 L of water. This represents 75% of the total capacity of the tank. How much water can the tank hold when full?

☐ 120 L   ☐ 270 L   ☐ 480 L   ☐ 600 L

SHADE ONE BOX

A bedroom has dimensions of 5 m × 4 m. The scale drawing of the bedroom is shown below.

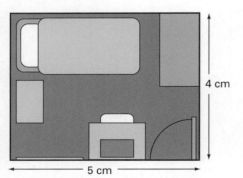

4 cm

5 cm

What scale has been used in the drawing?

☐ 1 m represents 1 cm

☐ 1 m represents 2 cm

☐ 1 cm represents 1 m

☐ 1 cm represents 2 m

QUESTION 23

A survey of average hours of exercise per day was carried out by 100 Year 7 students from Johnston Secondary College. The results are tabulated below.

Average daily exercise (hours)	0	1	2	3	4	5
Number of Students	9	23	49	35	14	7

Which of the following is the graph that correctly represents this data?

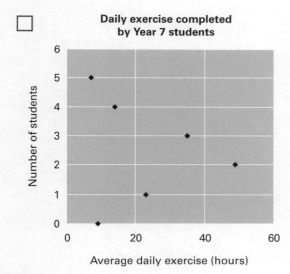

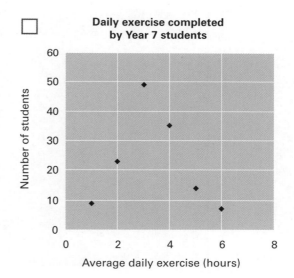

QUESTION 24

Hayden is the captain of a team of 8 players. The team must score 112 points to win a competition. To win the competition, what is the lowest average (mean) number of points that each player must score?

☐ 13   ☐ 14   ☐ 15   ☐ 16

QUESTION 25

What is the perimeter of this shape?

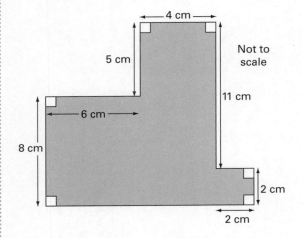

4 cm

5 cm

Not to scale

6 cm

8 cm

11 cm

2 cm

2 cm

☐ 38 cm   ☐ 46 cm   ☐ 50 cm   ☐ 67 cm

QUESTION 26

The graph below shows the average daily maximum temperature for Alice Springs, Northern Territory.

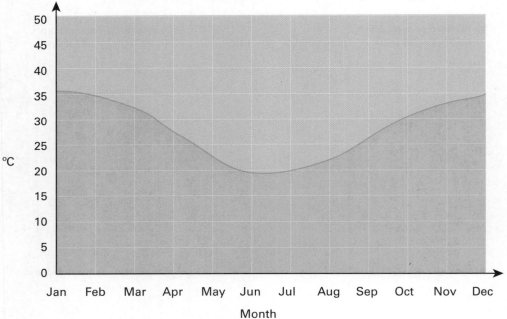

Mean daily maximum temperature at Alice Springs Airport

°C

Jan Feb Mar Apr May Jun Jul Aug Sep Oct Nov Dec

Month

(Source:weatherzone.com.au)

In which months did the average daily maximum temperature fall below 30°C?

☐ between May and August

☐ from January to March only

☐ between April and September

☐ from October to December only

QUESTION 27

This is a map of a nine-hole golf course.

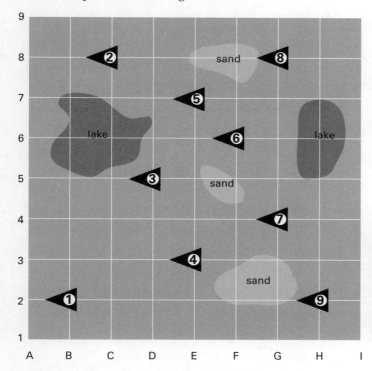

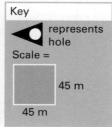

Which one of the following statements is correct?

☐ hole 9 is 225 m east of hole 1

☐ hole 4 is 180 m south of hole 5

☐ hole 7 is 135 m south of hole 8

☐ hole 8 is 160 m west of hole 2

QUESTION 28

 WRITE YOUR OWN ANSWER

Laura shares some money equally among 8 friends, with each person receiving $12. If Laura shares the same amount between 6 friends, how much money would each person receive?

$ [   ]

QUESTION 29

SHADE ONE BOX

This bar graph, drawn to scale, shows the favourite fruits of 160 customers at a fruit shop during a summer period.

Mango	Grapes	Berries	Melon	Peach

How many customers preferred berries?

☐ 16   ☐ 30   ☐ 32   ☐ 48

WRITE YOUR OWN ANSWER

Olivia uses these two conversion graphs:

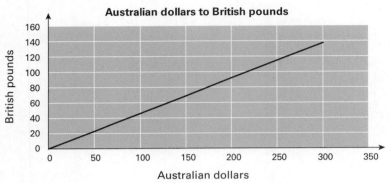

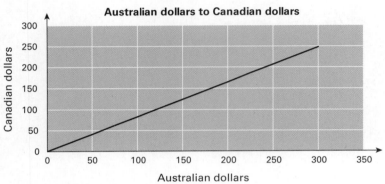

How many Canadian dollars are equal to 70 British pounds?

QUESTION 31

WRITE YOUR OWN ANSWER

This is an airport timetable for flights from Brisbane to other Australian cities:

Timetable of Flights		
Flight	Departure Times	Length of Flight
Brisbane to Canberra	8.30 a.m., 10.00 a.m., 1.30 p.m., 3.45 p.m.	1 h 50 min
Canberra to Adelaide	9.00 a.m., 11.15 a.m., 1.30 p.m., 5.30 p.m.	1 h 45 min
Adelaide to Brisbane	10.00 a.m., 1.45 p.m., 6.30 p.m., 7.15 p.m.	2 h 25 min

David lives in Brisbane and has a work meeting to attend today in Canberra. The meeting is at 11.00 a.m. in Canberra and will go for two hours. David will then fly to Adelaide for another meeting. This meeting will go for 2 h 30 min. What is the earliest time David will arrive back in Brisbane tonight?

p.m.

QUESTION 32

WRITE YOUR OWN ANSWER

Jennifer bought 240 g of chocolates. One quarter of the chocolates she bought contained nuts. White chocolate made up $\frac{1}{3}$ of the remaining chocolates. If Jennifer bought 96 g of dark chocolate, what fraction of the total amount of chocolate purchased was milk chocolate without nuts?

Test 6: Calculator allowed

Instructions

- A correct answer scores 1 mark, and an incorrect answer scores 0.
- Marks are not deducted for incorrect answers.
- No marks are given if more than one answer alternative is shaded.
- Choose the alternative which most correctly answers the question and shade in the box next to it.

QUESTION 1

SHADE ONE BOX

Which of the following pairs are equivalent?

☐ 33% and $\frac{1}{3}$ ☐ 0.5 and 5% ☐ 0.24 and $\frac{1}{4}$ ☐ $\frac{1}{5}$ and 0.2

QUESTION 2

SHADE ONE BOX

What is the area of the rectangle?

15.1 cm

8.2 cm

☐ 123.82 cm² ☐ 147.12 cm² ☐ 294.24 cm² ☐ 123.82 cm²

QUESTION 3

SHADE ONE BOX

A spinner is spun 10 000 times and the results are shown in the graph.

Results

Number of spins

10 000
9000
8000
7000
6000
5000
4000
3000
2000
1000
0

Which of the following spinners is most likely to give these results?

☐ Spinner 1 ☐ Spinner 2 ☐ Spinner 3 ☐ Spinner 4

QUESTION 4

SHADE ONE BOX

Charlie was making terrace houses from matchsticks. Once she had finished, she drew a graph of her results.

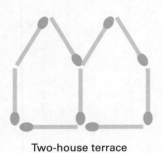

Two-house terrace

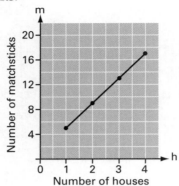

What is the rule that connects the number of houses (h) to the number of matchsticks (m)?

- [ ] h = 4m + 1
- [ ] h = m + 4
- [ ] m = 4h + 1
- [ ] m = h + 4

QUESTION 5

SHADE ONE BOX

Which one of the following statements is true?

- [ ] The highest common factor of 12 and 20 is 2.
- [ ] The highest common factor of 15 and 30 is 2.
- [ ] The highest common factor of 15 and 30 is 6.
- [ ] The highest common factor of 12 and 30 is 6.

QUESTION 6

SHADE ONE BOX

Four people went shopping at the supermarket. Their dockets are shown below. Who spent the most?

Kirstie	Scott	Brigid	Ronald
1.14	49.03	10.95	39.67
69.15	55.33	10.93	59.02
91.60	96.08	22.38	112.89
22.55	3.65	62.08	46.74
1.28	27.65	19.79	16.82
46.36	77.11	7.10	138.96
84.07	182.69	34.10	38.32
59.07	46.88	86.37	38.62
58.79	95.58	36.86	38.62
93.72	32.20	23.85	36.63

- [ ] Kirstie
- [ ] Scott
- [ ] Brigid
- [ ] Ronald

9780170462204

QUESTION 7

Which of the following solids below is identical to this one?

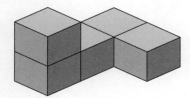

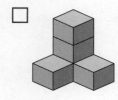 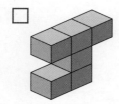 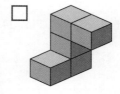 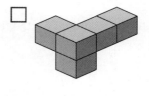

QUESTION 8

Mario is standing in the middle of a room, facing the window, as shown.

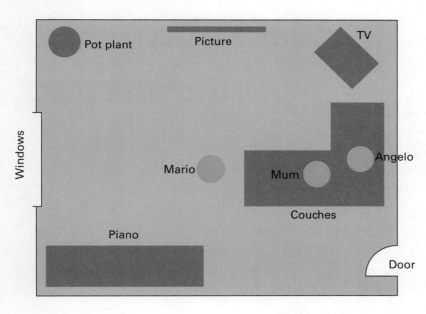

If Mario has turned anticlockwise to face the TV, through what type of angle has he turned?

- [ ] an acute angle
- [ ] an obtuse angle
- [ ] a right angle
- [ ] a reflex angle

QUESTION 9

WRITE YOUR OWN ANSWER

What is the sixth element in this pattern?

1, 2, 4, 8 …

QUESTION 10

SHADE ONE BOX

Which set of measurements is written in ascending order?

☐ 3.1 cm, 3.01 cm, 31.5 mm, 31.05 mm

☐ 31.5 mm, 31.05 mm, 3.01 cm, 3.1 cm

☐ 31.5 mm, 31.05 mm, 3.1 cm, 3.01 cm

☐ 3.01 cm, 3.1 cm, 31.05 mm, 31.5 mm

QUESTION 11

SHADE ONE BOX

The traditional date for the flight of Muhammad is AD 662.

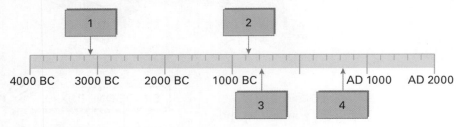

On which label on the time line would you place this event?

☐ label 1   ☐ label 2   ☐ label 3   ☐ label 4

QUESTION 12

 WRITE YOUR OWN ANSWER

A la Romana Pizza House has the following price list.

PIZZA				
	Small	**Medium**	**Large**	**Family**
BBQ Chicken	$5.50	$6.50	$7.50	$9.50
Ham & Cheese	$4.90	$5.90	$6.90	$8.90
Meatlovers	$4.90	$5.90	$6.90	$8.90
Vegetarian	$4.90	$5.90	$6.90	$8.90
The Lot	$4.90	$5.90	$6.90	$8.90
Cheese	$5.50	$6.50	$7.50	$9.50
Hawaiian	$4.90	$5.90	$6.90	$8.90
Mexicana	$4.90	$5.90	$6.90	$8.90
Extras	$0.40	$0.60	$0.80	$1.00
Garlic bread $2.20		Herb bread $2.20		
Delivery fee $2.50				

Damien and his friends order one medium meatlovers pizza, one garlic bread and two large cheese pizzas to be delivered. How much will it cost?

$ _____

QUESTION 13

SHADE ONE BOX

One measure of the difficulty of reading is the length of words in a random paragraph. The graph below shows the analysis of a paragraph in *The Dark Sun*.

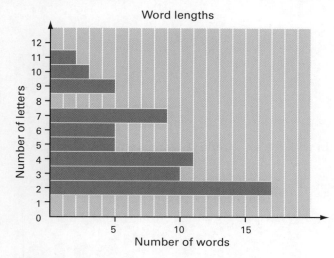

What statement below is true about this graph?

☐ The longest word had 15 letters.

☐ There was the same number of words with 5 and 6 letters.

☐ The most common number of letters in a word was 4.

☐ There were no words with 7 letters.

QUESTION 14

WRITE YOUR OWN ANSWER

What is the missing number?

$27 \times \boxed{} = 18 \times 18 \times 8$

QUESTION 15

SHADE ONE BOX

The grey shape in the diagram has been rotated to make the orange shape.

What is the best description of the rotation?

☐ 90° clockwise

☐ 90° anticlockwise

☐ 180° clockwise

☐ 270° clockwise

QUESTION 16

Nana's button tin has lots of white buttons and lots of coloured buttons.
This shows the ratio of white buttons to coloured buttons in Nana's tin.

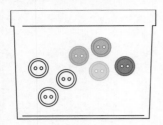

Which of the following diagrams shows another tin with the same ratio of white to coloured buttons?

☐   ☐   ☐   ☐

 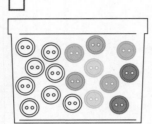 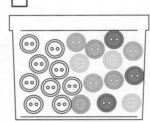 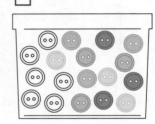

QUESTION 17

What is the rule shown by this flow chart?

☐ $m = 4n + 3$ ☐ $m = n + 7$ ☐ $n = m + 7$ ☐ $n = 4m + 3$

QUESTION 18

Which letter has the coordinates (3.5, 8.5)?

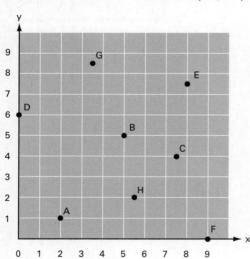

☐ G   ☐ H   ☐ E   ☐ C

QUESTION 19

SHADE ONE BOX

Which of the following statements is true about the following shapes?

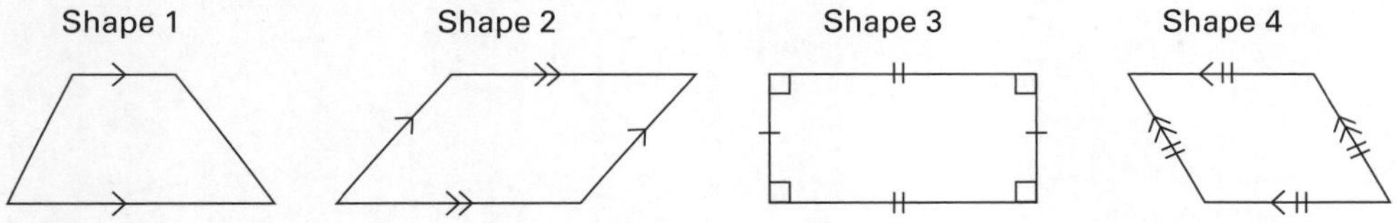

Shape 1　　　　Shape 2　　　　Shape 3　　　　Shape 4

☐ Shape 1 is a rhombus.

☐ Shape 2 is a trapezium.

☐ Shape 3 is a rectangle.

☐ Shape 4 is a square.

QUESTION 20

SHADE ONE BOX

Which of the following is the most likely answer for the area of a stamp?

☐ 2 mm^2　　　　☐ 2 cm^2　　　　☐ 2 m^2　　　　☐ 2 km^2

QUESTION 21

WRITE YOUR OWN ANSWER

The graph below shows the number of students in each class at Fairview Primary School.

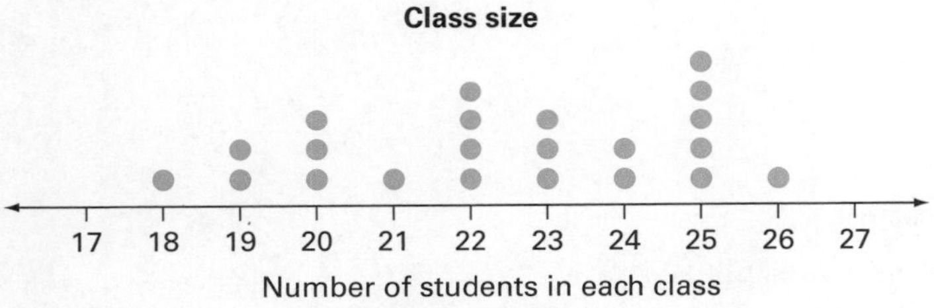

Class size

Number of students in each class

What is the range of the data?

QUESTION 22

SHADE ONE BOX

Which of the following diagrams has 25% shaded?

☐　　☐　　☐　　☐

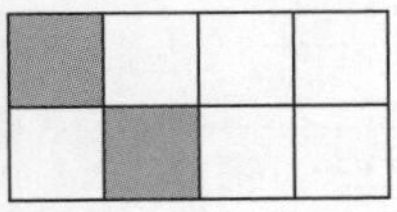

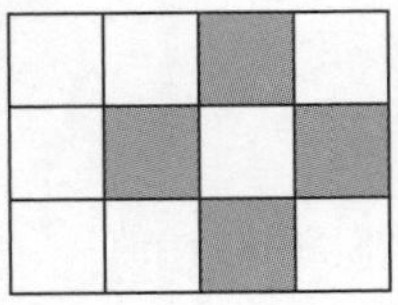

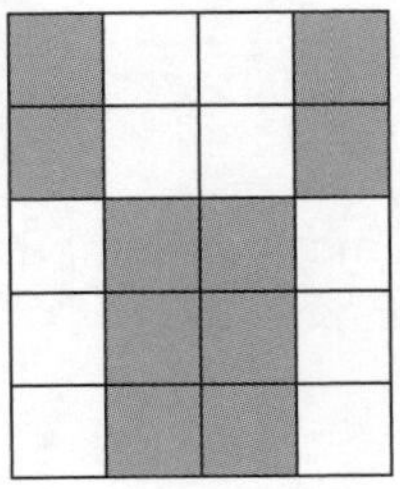

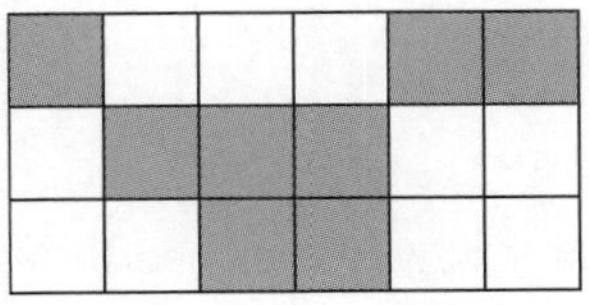

QUESTION 23

 WRITE YOUR OWN ANSWER

An L-shaped pattern is made with tiles as shown.

Arm length 1

Arm length 2

What is the value of x in the table?

Arm length	Number of tiles
1	3
2	5
3	
4	
5	x

QUESTION 24

SHADE ONE BOX

The table below shows the road distance between mainland Australian capital cities.

Adelaide						
2131	Brisbane					
1210	1295	Canberra				
3212	3493	4229	Darwin			
745	1736	655	3957	Melbourne		
2749	4390	3812	4342	3494	Perth	
1431	1027	304	4060	893	3988	Sydney

Which two cities are closest together by road?

☐ Melbourne and Canberra

☐ Brisbane and Sydney

☐ Canberra and Sydney

☐ Melbourne and Sydney

QUESTION 25

SHADE ONE BOX

Many newspaper headlines have a measurement theme. Which of the following headlines has a length theme?

☐ Ozone layer thinning!

☐ Nugget bigger than the 'Welcome Stranger' found by accident.

☐ Queensland's hottest day!

☐ Category 3 earthquake near Tonga.

9780170462204

QUESTION 26

What is the size of angle b in the diagram?

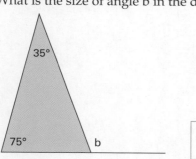

degrees

QUESTION 27

Tickets booked for a rock festival are charged according to the
following rule c = 75n + 6.50 where c is the total cost, n is the number of tickets, $75 is the ticket price and
$6.50 is the booking fee. How much would Mia, Rhea and Leah pay altogether if they buy tickets this way?

$

QUESTION 28

What is the volume of the solid below?

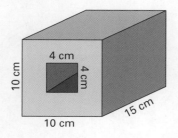

cm³

QUESTION 29

Below is part of a timetable for trains to the city.

Trains to the city							
Plymouth	6.05	6.10	6.27	6.42	7.00		7.16
Block		6.12	6.29		7.02		
Glamis		6.14	6.31		7.04		
Irwin	6.08	6.16	6.33		7.06	7.08	
Gardener	6.10	6.18	6.35		7.08	7.10	
Hodge	6.12	6.20	6.37		7.10	7.12	
Elder	6.14	6.22	6.39		7.12	7.14	
Brougham	6.16	6.24	6.41	6.51	7.14	7.16	7.25
Welly	6.19	6.27		6.54	7.17	7.19	7.28
City	6.22	6.30	6.45	6.57	7.20	7.22	7.31

How long does it take on the express train to get from Plymouth to Welly?

minutes

The brochure shows the cost of the phone company CallCheap.

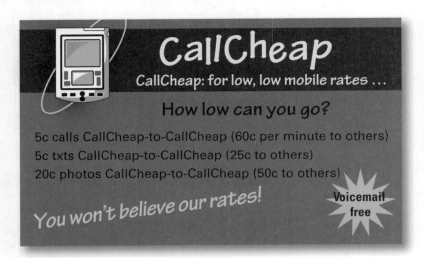

Tessa's friends all use this company except George. One day she sends 120 texts to her friends that use CallCheap and 20 to George. She makes 35 calls to her CallCheap friends and speaks to George for 5 minutes. She also sends George three photos. How much would she pay if she used CallCheap?

$ [   ]

QUESTION 31

SHADE ONE BOX

The following solid was made from cubes.

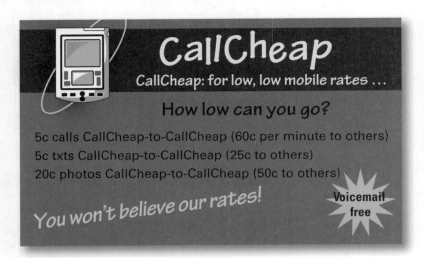

Assuming the cubes go down to the ground, the number of cubes required to make the solid is:

☐ 10   ☐ 12   ☐ 8   ☐ 11

QUESTION 32

SHADE ONE BOX

The ancient Greeks considered a number to be a **perfect number** if the sum of its factors, excluding itself, equals the number. Which of the following is a perfect number?

☐ 26   ☐ 25   ☐ 27   ☐ 28

Year 7 Numeracy
Non-calculator
Full-length Test 7

Writing time: 40 minutes

Use 2B pencil only

Instructions

- Write your **student name** in the space provided.
- You must be silent during the test.
- If you need to speak to the teacher, raise your hand. Do not speak to other students.
- Answer all questions using a 2B pencil.
- If you wish to change your answer, erase it thoroughly and then write your new answer.
- Students are NOT permitted to bring a calculator into the test room.

Student name:

QUESTION 1

WRITE YOUR OWN ANSWER

483 – 152 = ☐

QUESTION 2

Which dotted line is a line of symmetry?

SHADE ONE BOX

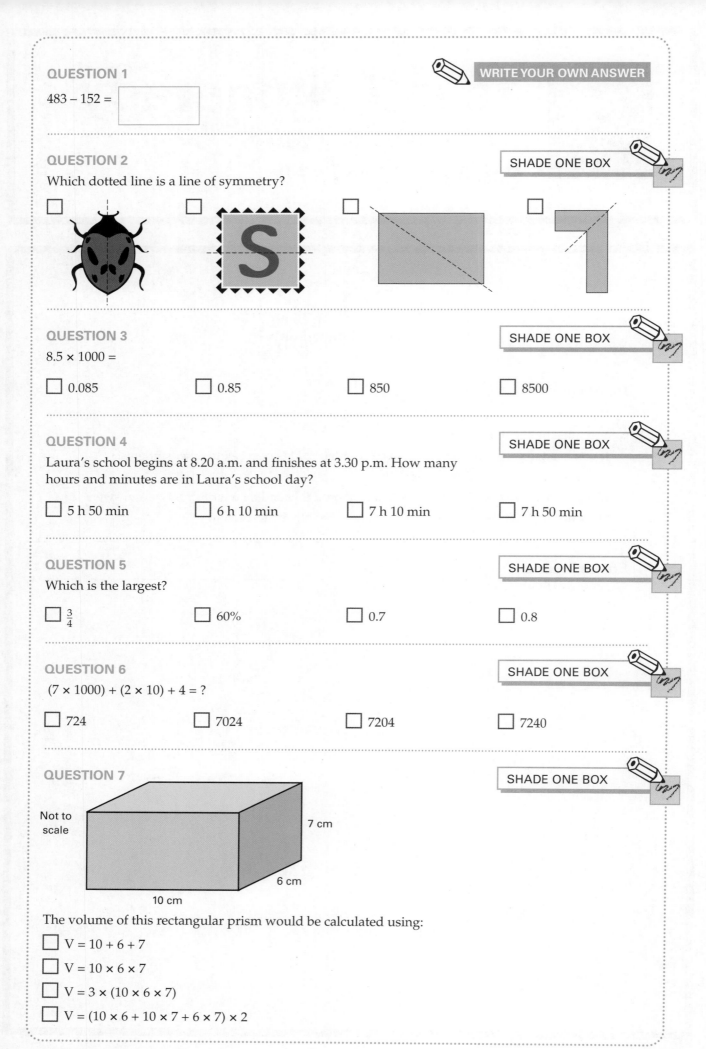

☐   ☐   ☐   ☐

QUESTION 3

SHADE ONE BOX

8.5 × 1000 =

☐ 0.085   ☐ 0.85   ☐ 850   ☐ 8500

QUESTION 4

SHADE ONE BOX

Laura's school begins at 8.20 a.m. and finishes at 3.30 p.m. How many hours and minutes are in Laura's school day?

☐ 5 h 50 min   ☐ 6 h 10 min   ☐ 7 h 10 min   ☐ 7 h 50 min

QUESTION 5

SHADE ONE BOX

Which is the largest?

☐ $\frac{3}{4}$ ☐ 60%   ☐ 0.7   ☐ 0.8

QUESTION 6

SHADE ONE BOX

(7 × 1000) + (2 × 10) + 4 = ?

☐ 724   ☐ 7024   ☐ 7204   ☐ 7240

QUESTION 7

SHADE ONE BOX

Not to scale

7 cm

6 cm

10 cm

The volume of this rectangular prism would be calculated using:

☐ V = 10 + 6 + 7

☐ V = 10 × 6 × 7

☐ V = 3 × (10 × 6 × 7)

☐ V = (10 × 6 + 10 × 7 + 6 × 7) × 2

QUESTION 8

Look at the column graph shown below.

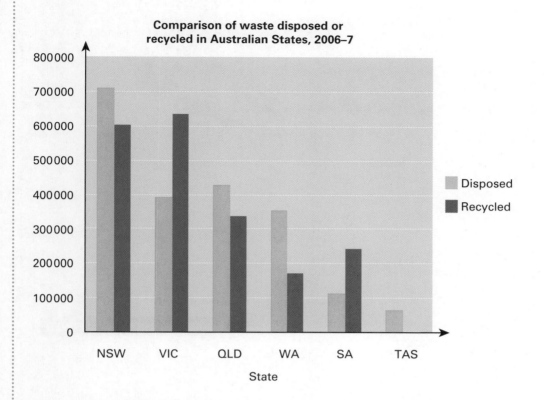

How many states recycled more materials than they disposed of?

☐ 1   ☐ 2   ☐ 3   ☐ 4

QUESTION 9

Which diagram shows an obtuse angle?

☐ 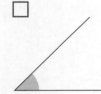   ☐ 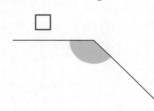   ☐ 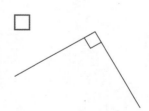   ☐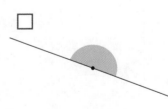

QUESTION 10

Owen has a bag of 12 marbles. If a marble is taken from the bag without looking, the probability of selecting a blue marble is $\frac{1}{3}$. How many blue marbles does Owen have?

☐ 2

☐ 3

☐ 4

☐ 6

What is the best way to estimate the total cost of these items?

beach towel
$29

backpack
$89

joggers
$63

☐ $30 + $80 + $60

☐ $30 + $80 + $70

☐ $30 + $90 + $60

☐ $30 + $90 + $70

QUESTION 12

WRITE YOUR OWN ANSWER

One quarter of Matthew's savings is $12. How much money does Matthew have in savings?

$ _____

QUESTION 13

WRITE YOUR OWN ANSWER

The graph shows the types of drinks preferred by children at a sports match.

Children's drink choices

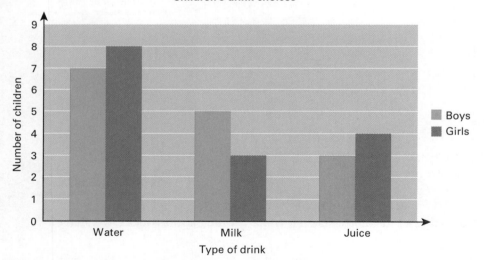

Use the information in the graph to complete this table.

Drinks	Boys	Girls
Water	7	8
Milk		3
Juice	3	4

QUESTION 14

SHADE ONE BOX

If $a = 6$ and $b = 4$, the correct substitution for ab is:

- ☐ 6×4
- ☐ $6 + 4$
- ☐ $6 - 4$
- ☐ $6 \div 4$

QUESTION 15

WRITE YOUR OWN ANSWER

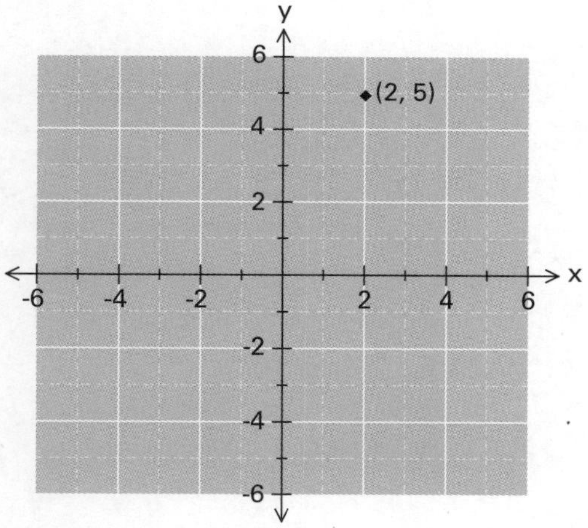

y
x
(2, 5)
6
4
2
-2
-4
-6

What is the ordered pair, 4 units to the left and 1 unit down from (2, 5)?

(,)

QUESTION 16

SHADE ONE BOX

A number line is shown below.

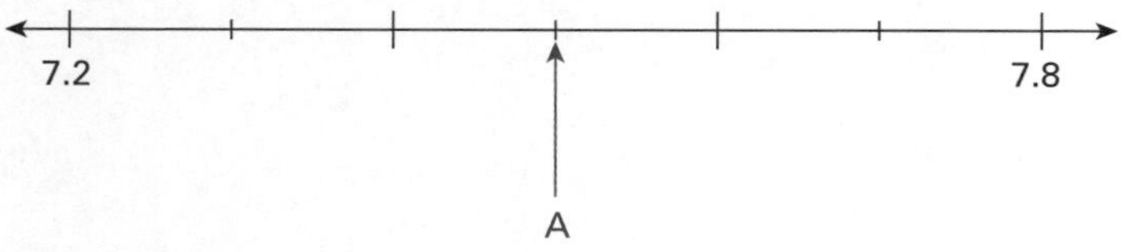

7.2
7.8
A

What number is represented on the number line by A?

- ☐ 7.3
- ☐ 7.4
- ☐ 7.5
- ☐ 7.6

QUESTION 17

SHADE ONE BOX

In the shape below, what is the total length of the two sides with dotted lines?

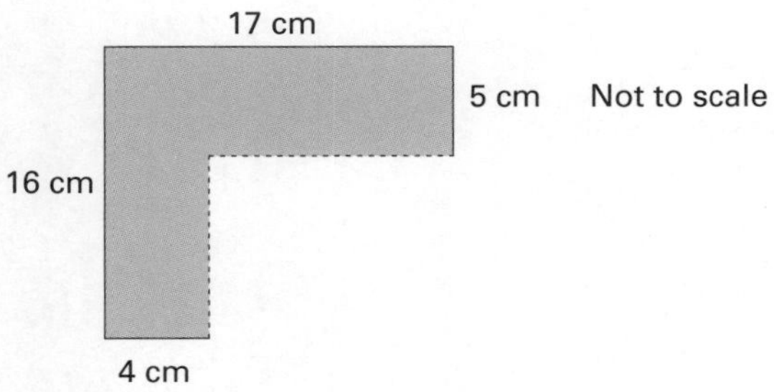

17 cm
5 cm
Not to scale
16 cm
4 cm

- ☐ 24 cm
- ☐ 22 cm
- ☐ 21 cm
- ☐ 17 cm

QUESTION 18

A flow chart is shown below.

 2 →(× 10)→ [ ] →(− 5)→ [ ] →(+ 7)→ [?]

What is the missing number?

[ ]

QUESTION 19

Ben owns a fruit shop. He constructed a line graph to show the price of one punnet of berries over a five-week period.

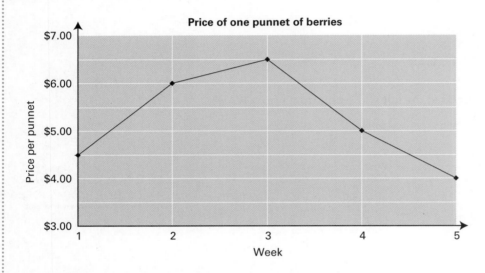

Price of one punnet of berries

Which of the following statements is true?

☐ In week 1, the cost of two punnets of berries was $10.

☐ The price of one punnet of berries was highest in week 3.

☐ The cost of one punnet of berries doubled in week 2.

☐ The largest drop in price for one punnet of berries was from week 4 to week 5.

QUESTION 20

Which solid has exactly six faces and 12 edges?

☐ 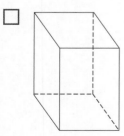   ☐ 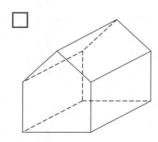   ☐ 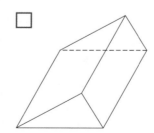   ☐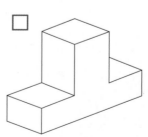

QUESTION 21

SHADE ONE BOX

The diagram below shows a rectangle with length 6 cm and width 4 cm.

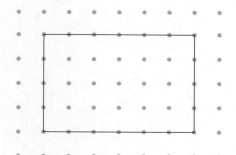

Which triangle has the same area as this rectangle?

☐ 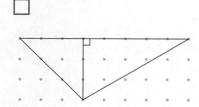   ☐    ☐   ☐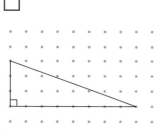

QUESTION 22

WRITE YOUR OWN ANSWER

The maximum and minimum daily temperatures (°C) in London for one week in February are shown below.

	Sunday	Monday	Tuesday	Wednesday	Thursday	Friday	Saturday
Minimum Temperature (°C)	-2	1	0	-1	-2	0	-1
Maximum Temperature (°C)	2	4	4	3	3	4	5

Which day shows the greatest difference in temperature?

QUESTION 23

SHADE ONE BOX

Which of the following is equivalent to $3x + 4xy + 5x + y$?

☐ $12x^2y^2$

☐ $y + 5xy$

☐ $8x + 5y$

☐ $8x + 4xy + y$

QUESTION 24

SHADE ONE BOX

What is the size of the angle marked x° in the triangle?

- ☐ 30°
- ☐ 45°
- ☐ 55°
- ☐ 60°

QUESTION 25

SHADE ONE BOX

Min and Scott share $30 between them in the ratio 2 to 1. How much money does Min receive?

- ☐ $20
- ☐ $15
- ☐ $10
- ☐ $5

QUESTION 26

WRITE YOUR OWN ANSWER

This pattern uses matchsticks to make squares.

This pattern is continued until there are 9 squares. How many matchsticks are used in total?

☐ matchsticks

QUESTION 27

SHADE ONE BOX

A car travels at a speed of 90 km/h. How far will it travel in 1 hour and 20 minutes?

- ☐ 75 km
- ☐ 100 km
- ☐ 108 km
- ☐ 120 km

QUESTION 28

SHADE ONE BOX

Sydney is two hours behind New Zealand in time. A plane leaves New Zealand at 9.00 a.m. and flies to Sydney. The flight takes three hours. What time is it in Sydney when the plane arrives?

- ☐ 8 a.m.
- ☐ 9 a.m.
- ☐ 10 a.m.
- ☐ 11 a.m.

9780170462204

WRITE YOUR OWN ANSWER

Luke uses these two conversion graphs.

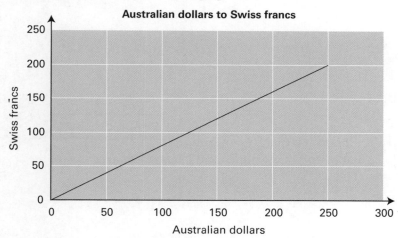

Australian dollars to Swiss francs

Australian dollars to Hong Kong dollars

How many Hong Kong Dollars are equal in value to 100 Swiss Francs?
Give your answer to the nearest $100.

Approx [　　　　] $HK

QUESTION 30

SHADE ONE BOX

Michael has an original photo with dimensions 6 cm by 8 cm. Michael
wants to enlarge the photo so it has a height of 20 cm.

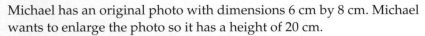

8 cm

6 cm

20 cm

?

What will be the width of the enlarged photo?

☐ 18 cm   ☐ 16 cm   ☐ 15 cm   ☐ 12 cm

QUESTION 31

WRITE YOUR OWN ANSWER

Amy has $9 more than Mia. In total, they have $47. How much money does Mia have?

$ ☐

QUESTION 32

SHADE ONE BOX

A bowl of fruit punch is made for a party. Fruit punch is made up of fruit juice, lemonade and fruit pieces. Fruit juice makes up $\frac{3}{4}$ of the punch and lemonade makes up $\frac{1}{5}$ of the punch. What fraction of the fruit punch is fruit pieces?

☐ $\frac{1}{20}$ ☐ $\frac{4}{9}$ ☐ $\frac{11}{20}$ ☐ $\frac{5}{9}$

Year 7 Numeracy
Calculator allowed
Full-length Test 8

Writing time: 40 minutes

Use 2B pencil only

Instructions

- Write your **student name** in the space provided.
- You must be silent during the test.
- If you need to speak to the teacher, raise your hand. Do not speak to other students.
- Answer all questions using a 2B pencil.
- If you wish to change your answer, erase it thoroughly and then write your new answer.
- Students are permitted to bring a calculator into the test room.

Student name:

QUESTION 1

SHADE ONE BOX

What is the value of 7 in 4.871?

☐ hundreds

☐ thousands

☐ hundredths

☐ thousandths

QUESTION 2

SHADE ONE BOX

Which of the following is equivalent to $\frac{3}{4}$?

☐ 75%   ☐ $\frac{16}{20}$ ☐ 0.6   ☐ $\frac{36}{45}$

QUESTION 3

WRITE YOUR OWN ANSWER

The value of $2^4 + 3^3 + 4^2$ is?

QUESTION 4

SHADE ONE BOX

A jug contains 1.4 L of milk. Three glasses of milk, each one holding 200 mL, are poured. Which arrow shows the amount of milk left in the jug?

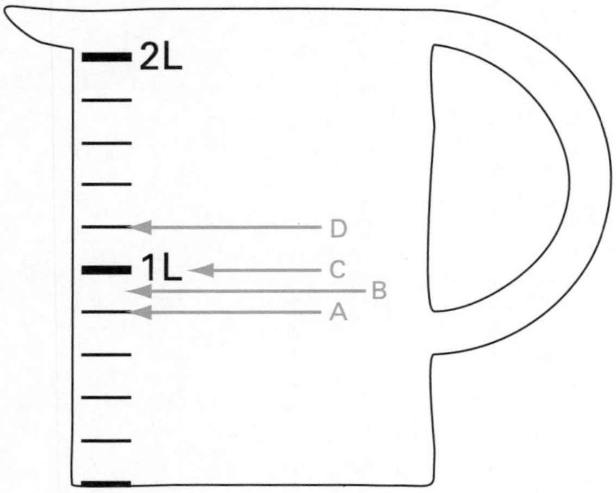

☐ A   ☐ B   ☐ C   ☐ D

QUESTION 5

SHADE ONE BOX

Chicken must be cooked 50 minutes for every kilogram it weighs. If a chicken weighs 3.5 kg, how long will it take to cook?

☐ 1 hour 55 minutes

☐ 1 hour 75 minutes

☐ 2 hours 45 minutes

☐ 2 hours 55 minutes

QUESTION 6

If $x = \frac{1}{2}$, what is the value of $6x + 8$?

QUESTION 7

Anne went to sleep at 8:42 p.m. and slept for 10 hours and 19 minutes. What time did Anne wake up?

QUESTION 8

This shape turns around the black dot.

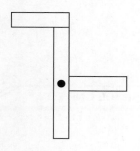

What does it look like after a 180 degree turn clockwise?

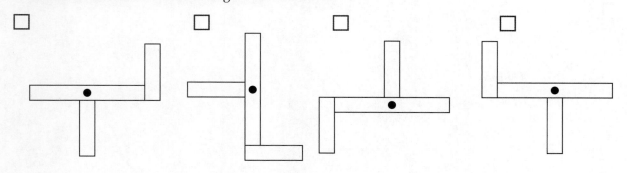

QUESTION 9

Which one of the following is an example of discrete data?

- the times run by students in a 100 m race
- the number of people attending a sporting match
- the birthplace of people living in Australia
- the heights of students in Year 7

QUESTION 10

Chloe makes a spinner. It is very likely, but not certain, that she will spin red.
Which spinner is Chloe's?

☐   ☐   ☐   ☐

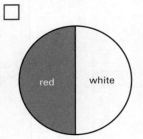 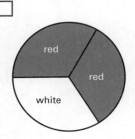 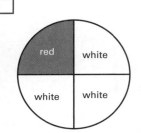 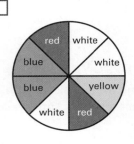

QUESTION 11

Calculate the size of ∠XYZ.

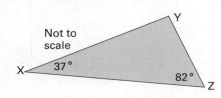

Not to scale

X 37° Y 82° Z

[   ] degrees

QUESTION 12

SHADE ONE BOX

This is a map of a zoo.

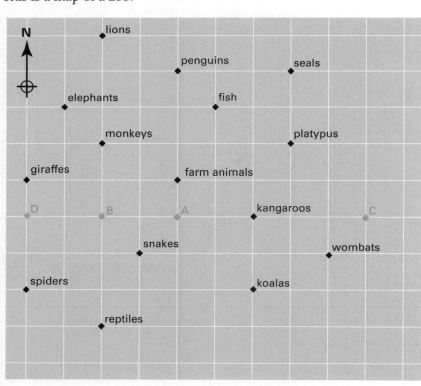

Marcus is standing due west of the kangaroos and south-east of the monkeys. Which point on the map shows where Marcus is standing?

☐ A   ☐ B   ☐ C   ☐ D

QUESTION 13

SHADE ONE BOX

Which one of the following events is best described as equally likely to occur?

☐ You toss a coin and it shows tails.

☐ You roll a normal die and you get a 1, 2, 3 or 4.

☐ A bag contains blue and red marbles and you choose, without looking, a yellow marble.

☐ You win first prize in a competition.

QUESTION 14

SHADE ONE BOX

Wendy has read 128 pages of a book containing a total of 466 pages. How many more pages does Wendy need to read to be halfway through the book?

☐ 64   ☐ 105   ☐ 233   ☐ 338

QUESTION 15

SHADE ONE BOX

Oliver has a cough and he takes cough medicine for it. His mother has a full 200 mL bottle of cough medicine. Each dose is 17 mL. How many full doses of cough medicine are contained in the bottle?

☐ 11   ☐ 12   ☐ 13   ☐ 14

QUESTION 16

SHADE ONE BOX

The number of ice creams sold during a heatwave was 450. Two flavours were sold to customers. The ratio of chocolate ice creams sold to strawberry ice creams sold was 3 to 2. How many strawberry ice creams were sold during the heatwave?

☐ 75   ☐ 180   ☐ 270   ☐ 300

QUESTION 17

SHADE ONE BOX

Which time is the same as the time shown on this digital clock?

17:30

☐ 5:30 a.m.

☐ 7:30 a.m.

☐ 5:30 p.m.

☐ 7:30 p.m.

QUESTION 18

The Johnson family has a monthly budget of $3750. How much do they spend each month on food?

[ ]

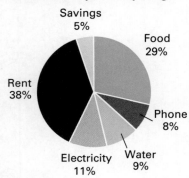

Johnson family monthly budget

Savings 5%
Food 29%
Rent 38%
Phone 8%
Water 9%
Electricity 11%

QUESTION 19

 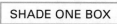

There were 24 students in class 7A. In a competition, the average number of points scored per student was 130. What was the total number of points scored by all the students in 7A?

[ ] 154   [ ] 2120   [ ] 2520   [ ] 3120

QUESTION 20

?

15 cm

The perimeter of a rectangle is 54 cm. If its length is 15 cm, what is its width?

[ ] 39 cm

[ ] 24 cm

[ ] 12 cm

[ ] 6 cm

QUESTION 21

Which one of these is equivalent to $12 + 4 \times 6 \div 3$?

[ ] $37 - 17 - 15 \times 4$

[ ] $60 - 20 \times 2 \div 4$

[ ] $9 \times 6 - 14 \div 2$

[ ] $56 \div (5 + 3) \times 3 - 1$

QUESTION 22

An object was made using identical cubes.

The diagram below shows the front view of the solid.

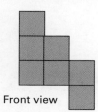

Front view

Which solid could have the front view shown above?

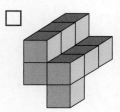 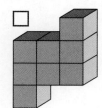 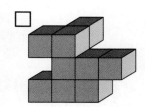 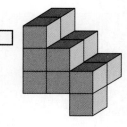

QUESTION 23

What is the missing number?

$3 \times \boxed{} - 15 = 18$

☐ 11   ☐ 9   ☐ 3   ☐ 1

QUESTION 24

Which number plane shows a set of ordered pairs satisfying the equation
$y = 3x - 2$?

☐
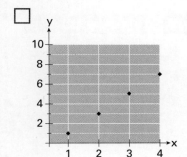

☐
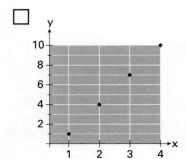

☐
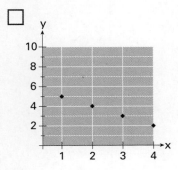

☐
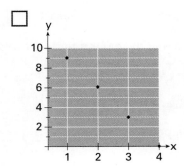

QUESTION 25

 WRITE YOUR OWN ANSWER

A piece of material 2.41 m long is cut into lengths of 0.06 m.
How many lengths of material can be cut?

lengths of material

QUESTION 26

 WRITE YOUR OWN ANSWER

Sarah is an artist. She paints four paintings for her friends as
presents. The paintings are all the same size, as indicated by the shape below.

50 cm

70 cm

Sarah wants to buy frames for each of the paintings. The material for each frame costs $6 per metre.
What will it cost Sarah to buy the material for the frames?

$ | |
|---|

QUESTION 27

SHADE ONE BOX

What is the angle sum of a quadrilateral?

☐ 90°   ☐ 180°   ☐ 360°   ☐ 450°

QUESTION 28

 WRITE YOUR OWN ANSWER

Miles is making this stack of cans for a supermarket display.

   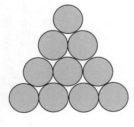

Shape 1   Shape 2   Shape 3   Shape 4

This table shows the number of cans used in the pattern for the display.

Shape	1	2	3	4	5
Number of cans	1	3	6	10	?

How many cans are needed to build shape 5?

cans

QUESTION 29

SHADE ONE BOX

A sale was advertised in a clothing shop window.

Sale – one week only!

15% discount on all items

A few days later, another sale was advertised in the clothing shop.

One-day sale!

Take another 20% off the sale price
of all items

William wants to buy a jacket. Its original price was $125. If William receives both discounts, what is the total amount William will save in the sale?

☐ $40   ☐ $43.75   ☐ $81.25   ☐ $85

QUESTION 30

SHADE ONE BOX

Here is a plan of Jack's backyard.

Key	
☐	Grass
▨	Flower bed

The area of the flower bed is 12 m². What is the area of the grass in Jack's backyard?

☐ 36 m²

☐ 48 m²

☐ 60 m²

☐ 72 m²

QUESTION 31

SHADE ONE BOX

The net of a die is shown below.

```
        ┌───┐
        │ 1 │
    ┌───┼───┼───┬───┐
    │ 2 │ 3 │ 2 │ 5 │
    └───┼───┼───┴───┘
        │ 4 │
        └───┘
```

If I roll the die once, what is the probability that I roll a 2?

☐ $\frac{1}{6}$ ☐ $\frac{1}{4}$ ☐ $\frac{1}{3}$ ☐ $\frac{1}{2}$

SHADE ONE BOX

QUESTION 32

A student scored the following marks, out of 20, in five weekly history tests:

Test 1: 18

Test 2: 15

Test 3: 11

Test 4: 15

Test 5: 16

Which one of the following statements is true?

- ☐ His average (mean) is 14.
- ☐ The mode of his marks is 15.
- ☐ The median of his marks is 11.
- ☐ His score in Test 3 has affected his median test mark.

9780170462204

Year 7 Numeracy
Non-calculator
Full-length Test 9

Writing time: 40 minutes

Use 2B pencil only

Instructions

· Write your **student name** in the space provided.
· You must be silent during the test.
· If you need to speak to the teacher, raise your hand. Do not speak to other students.
· Answer all questions using a 2B pencil.
· If you wish to change your answer, erase it thoroughly and then write your new answer.
· Students are NOT permitted to bring a calculator into the test room.

Student name:

QUESTION 1

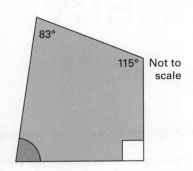

83°

115° Not to scale

What is the size of the angle marked in the quadrilateral?

QUESTION 2

18.7 km is equal to?

- [ ] 187 m
- [ ] 1870 m
- [ ] 18 700 m
- [ ] 187 000 m

QUESTION 3

The ages of five children are shown below.

Ki-Min: 10   Carlos: 7   Costa: 9   Monique: 10   Sasha: 4

What is the median age of these children?

- [ ] 7
- [ ] 8
- [ ] 9
- [ ] 10

QUESTION 4

$92 \times 7 = ?$

QUESTION 5

Which mixed numeral is equivalent to $\frac{37}{8}$?

☐ $3\frac{5}{8}$ ☐ $3\frac{7}{8}$ ☐ $4\frac{3}{8}$ ☐ $4\frac{5}{8}$

QUESTION 6

$0.902 = ?$

☐ $9 \times \frac{1}{10} + 2 \times \frac{1}{1000}$

☐ $9 \times \frac{1}{10} + 2 \times \frac{1}{100}$

☐ $9 \times 10 + 2 \times 1000$

☐ $9 \times 1 + 2 \times 100$

QUESTION 7

Which one of these events is most likely to occur?

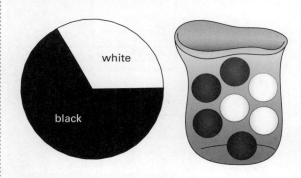

☐ the above spinner is spun once, landing on black

☐ without looking, choosing a black ball from the bag above

☐ tossing a coin and landing on heads

☐ from a bag containing 3 blue, 4 green and 5 yellow marbles, choosing, without looking, a blue marble

QUESTION 8

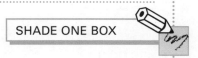

Which diagram shows $\frac{2}{3}$ of its area shaded?

☐   ☐   ☐   ☐

  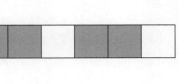   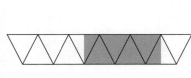

QUESTION 9

SHADE ONE BOX

If $636 \div 4 = 159$, then $6.36 \div 0.4 = ?$

☐ 0.159   ☐ 1.59   ☐ 15.9   ☐ 159

QUESTION 10

SHADE ONE BOX

Stefan has some diamond shaped tiles. Each edge of a tile is 4 cm long.

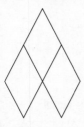

4 cm

Not to scale

He puts three tiles together to make this shape.

What is the perimeter of Stefan's shape?

☐ 40 cm   ☐ 36 cm   ☐ 32 cm   ☐ 28 cm

QUESTION 11

SHADE ONE BOX

What is $10 as a percentage of $50?

☐ 5%   ☐ 10%   ☐ 20%   ☐ 25%

QUESTION 12

SHADE ONE BOX

Which of the following solids is a prism?

☐ 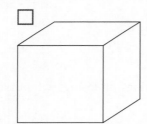   ☐   ☐ 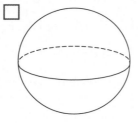   ☐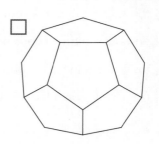

QUESTION 13

SHADE ONE BOX

If $m = 7$, then $2 \times (m - 3) = ?$

☐ 8   ☐ 10   ☐ 11   ☐ 24

9780170462204

QUESTION 14

SHADE ONE BOX

The shape below is turned about point P.

What will the shape look like after a half-turn clockwise about point P?

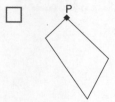 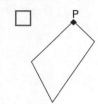 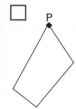 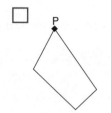

QUESTION 15

SHADE ONE BOX

This clock face shows the time a plane left Perth for Sydney one afternoon.

The plane trip took $5\frac{1}{2}$ hours. What time would the clock show when the plane arrives in Sydney?

  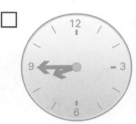 

   9:45 p.m.   8:30 p.m.   8:45 p.m.   9:15 p.m.

QUESTION 16

SHADE ONE BOX

These objects were made using identical cubes.

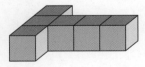

Which solid does **not** have the same volume as the solid above?

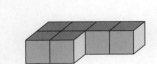 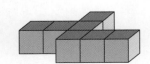 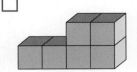 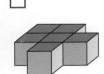

QUESTION 17

Which one of these is an equilateral triangle?

 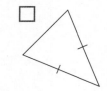 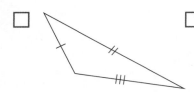 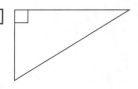

QUESTION 18

The daily rainfall for Hobart over seven days in March is shown below:

Day	Rainfall (mm)
Sunday	0 mm
Monday	2.6 mm
Tuesday	3.2 mm
Wednesday	16.4 mm
Thursday	2.2 mm
Friday	0 mm
Saturday	1.6 mm

What was the total rainfall for Hobart over this seven-day period?

millimetres

QUESTION 19

The below graph shows the distance Hasan travelled by car on a single journey.

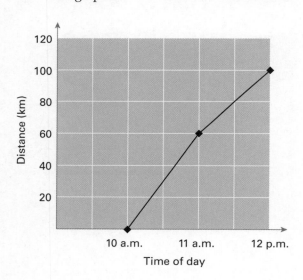

What was the speed of Hasan's car from 11 a.m. to 12 p.m.?

- [ ] 40 km/h
- [ ] 60 km/h
- [ ] 80 km/h
- [ ] 100 km/h

QUESTION 20

WRITE YOUR OWN ANSWER

Aisha scored the following marks out of 10 in three history tests.

Test 1	5
Test 2	8
Test 3	6
Test 4	?

What mark does Aisha need to score in Test 4 to have an average (mean) of 7 for the four tests?

QUESTION 21

SHADE ONE BOX

Here is a plan of a water fountain in a park.

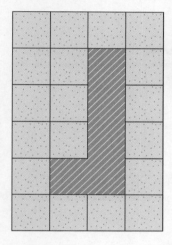

Key

☐ paving

▨ water

The area of the water fountain is 20 m². What is the area of the paving around the water fountain in the park?

☐ 4 m²   ☐ 76 m²   ☐ 19 m²   ☐ 8 m²

QUESTION 22

SHADE ONE BOX

The table below shows the skiing habits of students in a Year 7 class.

	Do ski	Do not ski
Male	14	3
Female	9	4

What is the probability that a female chosen at random from this class does not ski?

☐ $\frac{4}{9}$ ☐ $\frac{4}{13}$ ☐ $\frac{4}{7}$ ☐ $\frac{4}{30}$

QUESTION 23

SHADE ONE BOX

Simone uses this road map to travel from A to B. Each road can only be travelled in one direction as shown below.

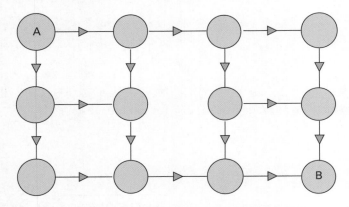

In how many ways can Simone travel from A to B?

☐ 6   ☐ 7   ☐ 8   ☐ 9

QUESTION 24

WRITE YOUR OWN ANSWER

I think of a number and subtract 3. I then multiply it by 3 and add 15.
The result is 42. What is the number?

QUESTION 25

SHADE ONE BOX

Look at the shape below.

How many triangles are in this shape?

☐ 20   ☐ 18   ☐ 14   ☐ 12

QUESTION 26

SHADE ONE BOX

The net of a rectangular prism is shown below.

5 m

2 m

10 m

What is the surface area of this rectangular prism?

☐ 170 m²   ☐ 140 m²   ☐ 100 m²   ☐ 90 m²

QUESTION 27

SHADE ONE BOX

Below is a map showing a number of ships near an island and a lighthouse.

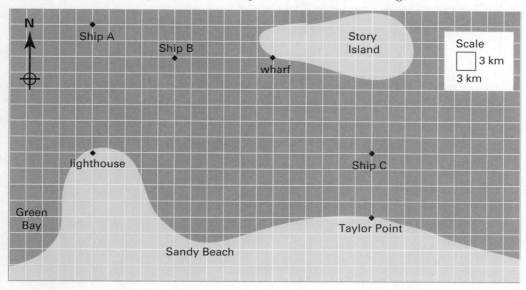

Which one of the following statements is correct?

- [ ] The lighthouse is 27 km due south of ship A.
- [ ] Ship B is 18 km west of the wharf.
- [ ] The lighthouse is 50 km due east of Ship C.
- [ ] Ship C is 15 km north of Taylor Point.

QUESTION 28

SHADE ONE BOX

If a = 1 and b = 2, find the value of $\frac{a}{2} + \frac{b}{3}$.

- [ ] $\frac{3}{5}$
- [ ] $\frac{5}{6}$
- [ ] $\frac{7}{5}$
- [ ] $\frac{7}{6}$

QUESTION 29

SHADE ONE BOX

The masses on the left-hand side of each of the scales below add to the amounts shown on the right-hand side of the scales.

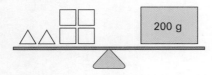 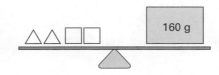

Which of the following combinations represents the weight of each object?

- [ ] △ = 40 g ☐ = 30 g
- [ ] △ = 20 g ☐ = 40 g
- [ ] △ = 60 g ☐ = 20 g
- [ ] △ = 70 g ☐ = 10 g

WRITE YOUR OWN ANSWER

A piece of metal pipe is cut in half. One half is used. One-third of the second half is cut off, it is 5 m long. What was the original length of the metal pipe?

| | metres
|---|

QUESTION 31

SHADE ONE BOX

Which table shows the points which satisfy the rule $y = 2x - 1$?

☐
x	1	3	6	7
y	1	5	11	13

☐
x	2	3	4	5
y	4	5	7	8

☐
x	2	4	6	8
y	3	7	10	14

☐
x	2	5	8	11
y	0	8	15	21

QUESTION 32

WRITE YOUR OWN ANSWER

Esther makes a rectangular-shaped birthday cake for her son. The cake is 20 cm long and 16 cm wide. Candles are placed every 4 cm apart and there is a candle in each corner.

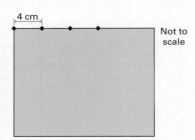

Not to scale

How many candles will Esther need on the birthday cake?

| | candles
|---|

Year 7 Numeracy
Calculator allowed
Full-length Test 10

Writing time: 40 minutes

Use 2B pencil only

Instructions

· Write your **student name** in the space provided.

· You must be silent during the test.

· If you need to speak to the teacher, raise your hand. Do not speak to other students.

· Answer all questions using a 2B pencil.

· If you wish to change your answer, erase it thoroughly and then write your new answer.

· Students are permitted to bring a calculator into the test room.

Student name:

QUESTION 1

SHADE ONE BOX

Which time is the same as that shown on this digital clock?

☐ 3.05 p.m.   ☐ 3.05 a.m.   ☐ 5.05 a.m.   ☐ 5.05 p.m.

QUESTION 2

SHADE ONE BOX

Another way of writing $5 \times 5 \times 5 \times 5 \times 5 \times 5$ is?

☐ 5^5 ☐ 5^6 ☐ 6^5 ☐ $5^2 \times 5^3$

QUESTION 3

SHADE ONE BOX

This is a hand-drawn map of Skull Island.

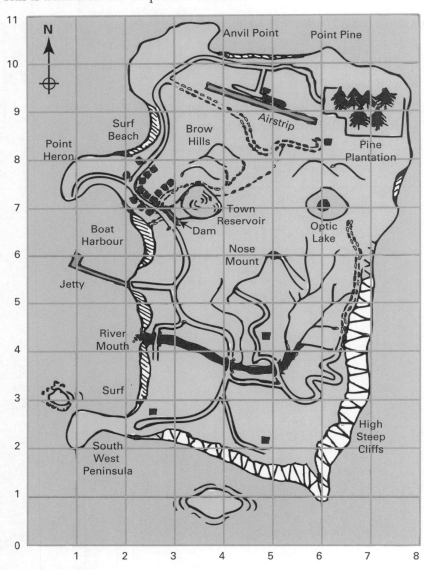

What feature is shown at (5, 6)

☐ Optic Lake   ☐ Pine Plantation   ☐ Jetty   ☐ Nose Mount

QUESTION 4

SHADE ONE BOX

Which of the following shapes is the same as the one below?

☐   ☐   ☐   ☐

QUESTION 5

SHADE ONE BOX

Which letter marks 4.4 cm?

☐ K   ☐ P   ☐ L   ☐ C

QUESTION 6

WRITE YOUR OWN ANSWER

What digit is represented by ● in the following?

```
   384
 − 1●3
  ─────
   231
```

QUESTION 7

SHADE ONE BOX

John has black and white socks. Getting dressed in the dark, he reaches into the sock drawer. He has an even chance of getting a black sock. Which of the following best represents his sock drawer?

☐   ☐   ☐   ☐

QUESTION 8

WRITE YOUR OWN ANSWER

Alyssa has made the following pattern with counters.

Term 1 Term 2 Term 3

How many counters will she put in the pattern that makes up Term 6?

QUESTION 9

SHADE ONE BOX

Calculate the change from $50 if you buy a cooked chicken at $9.36, 2 dozen eggs at $3.45 per dozen and 6.5 kg of apples at $2.50 per kilogram. Imagine there are still one- and two-cent coins.

☐ $32.51   ☐ $28.59   ☐ $18.69   ☐ $17.49

QUESTION 10

SHADE ONE BOX

Liam is a florist and he is making bouquets for the bridesmaids at a wedding. The ratio of white roses to pink roses in each bouquet must be 1:5. If he has five white roses in each bouquet, how many roses does he need altogether?

☐ 25   ☐ 30   ☐ 35   ☐ 20

QUESTION 11

SHADE ONE BOX

A swimming pool is 12 m by 3.8 m by 4 m. One litre of water occupies 0.001 m³. What is the capacity of the pool?

☐ 182.4 L
☐ 0.1824 L
☐ 18 240 L
☐ 182 400 L

QUESTION 12

SHADE ONE BOX

What is the best way to estimate the total cost of these three objects?

$44

$55

$39

☐ $30 + $50 + $40   ☐ $40 + $50 + $40   ☐ $40 + $60 + $40   ☐ $30 + $50 + $50

9780170462204

This graph shows the music preferences of some Year 7 students.

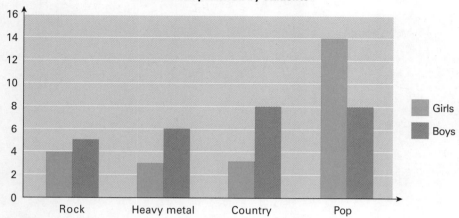

Music most preferred by students

Use the information in the graph to complete this table.

	Rock	Heavy metal	Country	Pop
Girls	4	3		14
Boys	5	6	8	8

QUESTION 14

SHADE ONE BOX

What of the following shows an angle of approximately 65°?

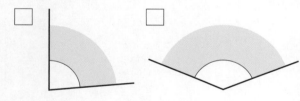 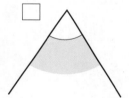 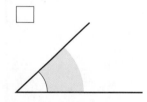

QUESTION 15

SHADE ONE BOX

Preety rode her bike at a speed of 6 m per second for a total of one-and-a-half minutes. How far did she ride?

☐ 9 m
☐ 540 m
☐ 720 m
☐ 480 m

Which of the following graphs represents y = 3x – 4?

 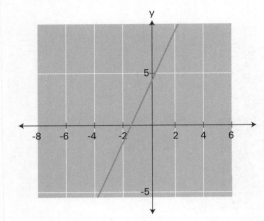

 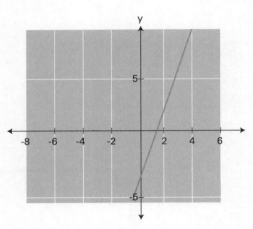

 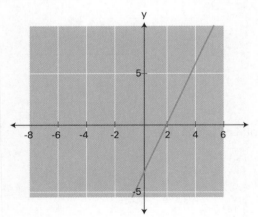

 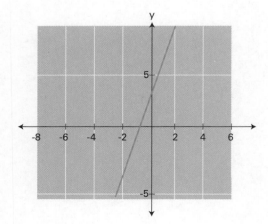

QUESTION 17

One-sixth of Gemma's scarf is 45 cm. How long is her scarf?
Give your answer in metres.

metres

QUESTION 18

A machine serves random drinks.

What is the probability that you will be served a Vampire blood drink from this machine?

☐ 25%   ☐ 50%   ☐ 75%   ☐ 100%

QUESTION 19

What is the angle and direction of rotation required to make L the image of H?

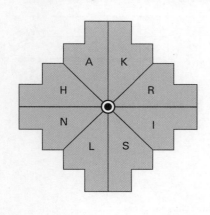

☐ 90° clockwise   ☐ 90° anticlockwise   ☐ 270° anticlockwise   ☐ 180° clockwise

QUESTION 20

Which of the following equations is represented by this flowchart?

$$x \quad -3 \quad \square \quad \times 4 \quad \square \quad \div 2 \quad \square \quad +5 \quad y$$

☐ $y = \dfrac{4(x-3)}{2} + 5$

☐ $x = \dfrac{4y-3}{2} + 5$

☐ $x = \dfrac{4(y-3)}{2} + 5$

☐ $y = \dfrac{4x-3}{2} + 5$

QUESTION 21

The ratio of girls to boys at a concert is 5:4. If there are 240 boys at the concert, how many attended altogether?

☐ 480   ☐ 432   ☐ 300   ☐ 540

QUESTION 22

Josh flew to Sydney from Hobart. His flight left at 10.37 a.m. and landed in Sydney after being delayed for several minutes at 3.41 p.m. How many minutes was he in the air?

☐ minutes

QUESTION 23

A fair coin is tossed 100 times. Four possible results of this experiment are listed below. Which of these results is most likely to occur?

☐ Heads occurs exactly 12 times.

☐ Heads occurs exactly 70 times.

☐ Heads occurs exactly 54 times.

☐ Heads occurs exactly 101 times.

QUESTION 24

Which of the following events would require an estimate of mass?

☐ George is driving from Dubbo to Broken Hill.

☐ Hans is comparing his grape production from this year to last year.

☐ Cara is working out whether to put the air conditioner on.

☐ Rohini is going to purchase water to fill her swimming pool.

Work out $2\frac{3}{5} - 1\frac{2}{3}$ Write your answer in simplest terms.

WRITE YOUR OWN ANSWER

QUESTION 26

The figure below is reflected in two mirrors as shown.

SHADE ONE BOX

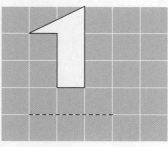

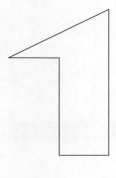

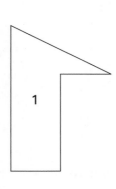

1

Mirror B

Mirror A

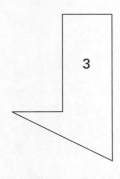

3

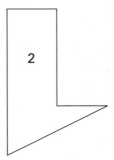

2

Which of the following statements is true?

☐ Image 3 is a reflection of Image 1 in mirror B.

☐ Image 3 is a reflection of Image 2 in mirror B.

☐ Image 2 is a reflection of Image 1 in mirror B.

☐ Image 2 is a reflection of the original in mirror B.

WRITE YOUR OWN ANSWER

Figure ABCD is enlarged by a scale factor to make figure JKLM.
What is the length of side JK, marked as x? Write your answer correct to one decimal place.

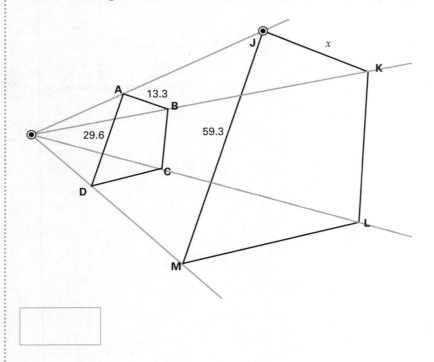

QUESTION 28

WRITE YOUR OWN ANSWER

Marcia sorted the lollies in her pack into colours and drew the following pie chart.

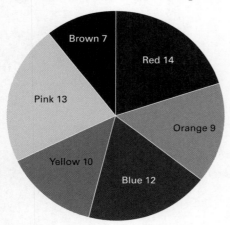

She only likes the yellow and orange lollies. What percentage of the box will she like? Write your answer correct to a whole number.

%

QUESTION 29

WRITE YOUR OWN ANSWER

For professional diving events, such as the Olympic Games, seven judges score a competitor out of a possible 10 points. Then the lowest and highest scores are rejected, and the mean of the remaining five scores is calculated to give the final score. For a particular competitor, the scores are: 6, 7, 8, 6, 8, 9 and 6.

What final score is awarded to this competitor?

QUESTION 30

SHADE ONE BOX

The heights of four students are 162.5 cm, 160.3 cm, 157.2 cm and 154.7 cm.

Use the clues below to find Ruby's height:

- Ruby is taller than Priscilla but shorter than Alice.

- Priscilla is shorter than Ai-Thie.

- Alice is not the tallest.

☐ 162.5 cm   ☐ 160.3 cm   ☐ 157.2 cm   ☐ 154.7 cm

QUESTION 31

 WRITE YOUR OWN ANSWER

Chris is making a pattern of triangles with matchsticks as shown below.

Write a rule that connects the number of triangles (t) with the number of matchsticks (m).

QUESTION 32

 WRITE YOUR OWN ANSWER

Find the surface area of this triangular prism.

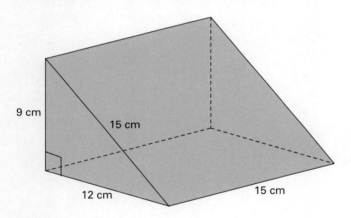

9 cm

15 cm

12 cm   15 cm

 cm²

Notes

9780170462204

Notes

Notes

9780170462204